UNCHARTED

UNCHARTED

Big Data as a Lens on Human Culture

EREZ AIDEN AND

JEAN-BAPTISTE MICHEL

RIVERHEAD BOOKS

a member of Penguin Group (USA)

New York

2013

RIVERHEAD BOOKS
Published by the Penguin Group
Penguin Group (USA) LLC
375 Hudson Street
New York, New York 10014

USA · Canada · UK · Ireland · Australia
New Zealand · India · South Africa · China

penguin.com
A Penguin Random House Company

ISBN 978-1-59448-745-3

Printed in the United States of America
1 3 5 7 9 10 8 6 4 2

BOOK DESIGN BY AMANDA DEWEY

For Aba,
who always believed I could count

EREZ AIDEN

To my family

JEAN-BAPTISTE MICHEL

CONTENTS

1

THROUGH THE LOOKING GLASS

Imagine if we had a robot that could read every book on every shelf of every major library, all over the world. It would read these books at a super-fast robot speed and remember every single word that it had read, using its super-infallible robot memory. What could we learn from this robot historian?

Here's a simple example that's familiar to every American. Today, we say that the southern states *are* full of southerners. We say that the northern states *are* full of northerners. We say that the New England states *are* full of New Englanders. Yet we say that the United States *is* full of citizens.

Why do we use the singular? This is more than a fine point of grammar: It's a matter of our national identity.

When the United States of America was established, its founding document, the Articles of Confederation, defined a weak central government, and referred to the new entity not as a single nation but instead as a "league of friendship" between individual states, somewhat akin to today's European Union. People thought of themselves not as Americans but as citizens of a particular state.

As such, citizens referred to "the United States" in the plural, as would be appropriate for a collection of distinct, mostly independent states. For instance, in President John Adams' 1799 State of the Union address, he talked about "the United States in *their* treaties with His Britannic Majesty." For a president to do that today would be inconceivable.

When did "We the People" (Constitution, adopted 1787) truly become "one nation" (Pledge of Allegiance, adopted 1942)?

If we asked human historians, they would probably point us to the most famous answer, from the end of James McPherson's celebrated Civil War history, *Battle Cry of Freedom*:

> . . . Certain large consequences of the war seem clear. Secession and slavery were killed, never to be revived during the century and a quarter since Appomattox. These results signified a broader transformation of American society and polity punctuated if not alone achieved by the war. Before 1861 the two words "United States" were generally rendered as a plural noun: "the United States are a republic." The war marked a transition of the United States to a singular noun.

McPherson wasn't the first to make this suggestion; this old chestnut has been discussed for at least a hundred years. Consider the following excerpt from the *Washington Post* in 1887:

> There was a time a few years ago when the United States was spoken of in the plural number. Men said "the United States are"—"the United States have"—"the United States were." But

> the war changed all that. Along the line of fire from the Chesapeake to Sabine Pass was settled forever the question of grammar. Not Wells, or Green, or Lindley Murray decided it, but the sabers of Sheridan, the muskets of Sherman, the artillery of Grant. . . . The surrender of Mr. Davis and Gen. Lee meant a transition from the plural to the singular.

Even a century later, it's hard not to get a thrill just reading this stirring tale of language, artillery, and adventure. Who could have dreamed of a war about grammar, or a subtle point of usage settled by "the muskets of Sherman"?

But should we believe it?

Probably. James McPherson is a former president of the American Historical Association and a legend among historians. *Battle Cry of Freedom*, his most famous work, won the Pulitzer. Moreover, whoever wrote that 1887 *Washington Post* article probably experienced this syntactic turnabout firsthand, and their eyewitness testimony couldn't be clearer.

Still, James McPherson, though brilliant, isn't infallible. And eyewitnesses sometimes get the facts wrong. Is there some way that we can do better?

Perhaps. Suppose that we ask our robot—the hypothetical robot that has read all the books in all the libraries—to contribute its mechanized opinion.

Suppose that, in response to our question, our helpful robot historian draws on its prodigious memory to make the chart that follows. The robot's chart shows how frequently the phrases "The United States is" and "The United States are" were used over time, in English books published in the United States. Horizontally, we see the flow of time, year by year. The vertical axis shows the frequency of the two phrases: how often they appear, on average, in every billion words of text written during the year in question. For instance, the robot read 313,388,047 words that appeared in books published in the year 1831. Within those words, the robot sees the phrase "The United States is" 62,759 times. That averages out to twenty times per billion words that year, indicated by the height of the corresponding line in 1831.

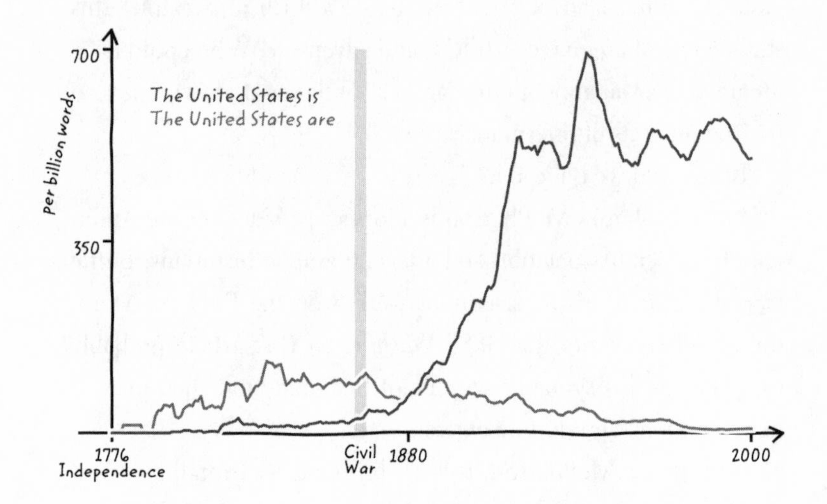

A chart like this would make it completely clear when people started talking about the United States in the singular.

There's just one small hitch: According to the hypothetical robot's hypothetical chart, the story we were telling you before is wrong. For one thing, the transition from plural to singular was not instantaneous. It was gradual, starting in the 1810s and continuing into the 1980s—a span of more than a century and a half. More important, there was no sudden switch during the Civil War. In fact, the war years did not differ much from the years immediately before or after. There was some postbellum acceleration, but it began five years after General Lee's surrender. According to the robot, the singular form did not become more common until 1880, fifteen years after the war. And even today, the plural banner of the state-spangled confederacy yet waves.

Of course, this is all hypothetical, because this stuff about a speed-reading robot outwitting an eyewitness and a prizewinning historian is so utterly far-fetched.

Except that it's all true.

McPherson, though brilliant, was wrong about the singular form. The eyewitness didn't recall events accurately. And the robot we were telling you about exists. And the chart we just showed you is the chart the robot drew. And there are a billion more charts it's just waiting to draw. And today, all over the world, millions of people are seeing history in a new way: through the digital eyes of a robot.

THE SHAPE OF THE LIGHT

This is not the first time that a new kind of lens has influenced how we look at the world.

In the late thirteenth century, a new invention, eyeglasses, began spreading like wildfire through Italy. In a matter of decades, glasses went from nonexistent to merely exotic to utterly commonplace. Forerunners of the smartphone, eyeglasses were an indispensable appliance for many Italians, combining fashion and function into an early triumph of wearable technology.

As eyeglasses spread across Europe and around the world, optometry became big business, and the technology for making lenses got better and cheaper. Inevitably, people began to experiment with what could be done when multiple lenses were combined. It wasn't long before folks realized that with a little bit of engineering, they could achieve extreme magnification. Compound lenses could be made to reveal new worlds invisible to the naked eye.

For instance, a compound lens could be used to magnify very small things. Microscopes uncovered at least two astonishing facts about the age-old mystery of life. They showed that the animals and plants all around us are subdivided into tiny, physically separate units. Robert Hooke, who made this discovery, noted that the arrangement of these units resembled the living quarters in monasteries, which is why he called them cells. Microscopes also revealed the existence of microbes. This separate universe of organisms, often made up of only a single cell, constitutes the vast majority of the living world. Prior to the invention of the microscope, no one had any idea that such life-forms might exist.

A compound lens could also be used to magnify faraway things. Armed with a telescope capable of 30X magnification—by modern standards, a child's plaything—Galileo tackled the mysteries of the cosmos. Wherever he looked, his telescope enabled him to see more than had ever been seen before. Pointing it at the moon—long believed to be a perfect sphere—the Florentine scientist saw valleys, plains, and mountains, the latter with distinct shadows that always pointed away from the sun. Exploring the bright band across the night sky called the Milky Way, Galileo could see that it consisted of stars, faint and innumerable: what today we call a galaxy. But Galileo's most famous discoveries came when he pointed his telescope at the planets. There he saw the phases of Venus and the moons of Jupiter, new worlds in the most literal sense.

Galileo's observations served as decisive evidence against the

Ptolemaic notion that the Earth stood still at the center of all things. Instead, they ushered in the Copernican view of the solar system: a sun surrounded by spinning planets. In Galileo's nimble hands, the optic lens—a mere trick of the light—both launched the scientific revolution and transformed the role of religion in Western life. It was more than the birth of modern astronomy. It was the birth of the modern world.

Even today, half a millennium later, the microscope and the telescope remain enormously relevant to the progress of science. Of course, the devices themselves have changed. Traditional optical imaging has become much more sophisticated, and some contemporary microscopes and telescopes rely on markedly different scientific principles. For instance, the scanning tunneling microscope uses ideas from twentieth-century quantum mechanics. Nonetheless, the scope of many sciences—in fields as diverse as astronomy, biology, chemistry, and physics—is still defined largely by their actual scopes—by what can be learned about those fields using the very best microscopes and telescopes available.

In 2005, when the two of us were graduate students, we spent a lot of time thinking about the kinds of scopes scientists had access to and the ways in which those scopes made science possible. We became intrigued by what seemed like an off-the-wall idea. For a long time, both of us had been interested in the study of history. We were especially fascinated by how human culture changes over time. Some of these changes are dramatic, but often they are so subtle as to be largely invisible to the unaided brain. Wouldn't it be great, we thought, if we had something like a microscope to measure human culture, to identify and track all those tiny effects that we would never notice otherwise? Or a telescope that would allow us to do this from a great distance—on other continents, centuries ago? In short, was it possible to create a kind of scope that, instead of observing physical objects, would observe historical change?

Of course, this would not be a Galileo-caliber contribution. The modern world already exists; the sun is already at the center of the solar system, and so on and so forth. Basically, everyone already knows that scopes are a good thing. But, we reasoned, this new kind of scope would probably be cool enough that Harvard might finally let us graduate, which is about all you can hope for when you're as underfed, underpaid, and overeducated as the typical PhD seeker.

As we were mulling this somewhat esoteric question, a revolution was occurring elsewhere that would sweep us up in its wake and lead millions of people to share our strange fascination. At its core, this big data revolution is about how humans create and preserve a historical record of their activities. Its consequences will transform how we look at ourselves. It will enable the creation of new scopes that make it possible for our society to more effectively probe its own nature. Big data is going to change the humanities, transform the social sciences, and renegotiate the relationship between the world of commerce and the ivory tower. To better understand how all this came about, let's take a close look at the historical record, from its modest beginnings to its omnipresent present.

COUNTING SHEEP

Ten thousand years ago, prehistoric shepherds periodically lost their sheep. Taking advice from prehistoric insomniacs, they hit on the idea of counting. Those very first accountants used stones as sheep counters, the same way that gamblers now use poker chips to keep track of their winnings.

All this worked very well. Over the next four thousand years, as people sought to track an increasingly wide array of goods, they used a simple carving instrument called a stylus to engrave patterns on some of the stones. These patterns could be used to indicate the different types of objects being counted. Eventually, in the fourth millennium BCE, someone decided that keeping track of a lot of little rocks—the Stone Age ancestors of loose change—was inconvenient. Instead, it was easier to take one really big stone and use the stylus to engrave lots of patterns on it, side by side. Writing was born.

In retrospect, it might seem surprising that something as mundane as the desire to count sheep was the impetus for an advance as fundamental as written language. But the desire for written records has always accompanied economic activity, since transactions are meaningless unless you can clearly keep track of who owns what. As such, early human writing is dominated by wheeling and dealing: a menagerie of bets, chits, and contracts. Long before we had the writings of the prophets, we had the writings of the profits. In fact, many civilizations never got to the stage of recording and leaving behind the kinds of great literary works that we often associate with the history of culture. What survives these ancient societies is, for the most part, a pile of receipts. If it weren't for the commercial enterprises that produced those records, we would know far, far less about the cultures that they came from.

This state of affairs is truer today than ever before. Unlike their predecessors, many of today's commercial enterprises do not create records as a mere by-product of doing business. Companies like Google, Facebook, and Amazon create tools that enable their users to represent themselves, and to interact with one another, on the Internet. These tools work by building a digital, personal, historical record.

For such companies, recording human culture *is* their core business.

And it's not just a record of things that were meant for public consumption, like Web pages, blogs, and online news. Increasingly, our personal communication, whether via e-mail, Skype, or text message, happens online. A lot of it is preserved there in some form, often by multiple entities, and in principle forever. Whether on Twitter or LinkedIn, both our personal and business relationships are enumerated on, and mediated by, the Web. When we "plus," "recommend," or send an e-card, our fleeting thoughts and impressions leave a permanent digital fingerprint. Google will remember every word of that angry e-mail long after we've forgotten the name of the person we sent it to. Facebook's photos will chronicle the details of that night at the bar even if we woke up with a fuzzy brain and a massive hangover. If we write a book, Google scans it; if we take a photo, Flickr stores it; if we make a movie, YouTube streams it.

As we experience all that contemporary life has to offer, as we live out more and more of our lives on the Internet, we've begun to leave an increasingly exhaustive trail of digital bread crumbs: a personal historical record of astonishing breadth and depth.

BIG DATA

How much information does all this add up to?

In computer science, the unit used to measure information is the bit, short for "binary digit." You can think about a single bit as the answer to a yes-or-no question, where 1 is yes and 0 is no. Eight bits is called a byte.

Right now, the average person's data footprint—the annual amount of data produced worldwide, per capita—is just a little short of one terabyte. That's equivalent to about eight trillion yes-or-no questions. As a collective, that means humanity produces five zettabytes of data every year: 40,000,000,000,000,000,000,000 (forty sextillion) bits.

Such large numbers are hard to fathom, so let's try to make things a bit more concrete. If you wrote out the information contained in one megabyte by hand, the resulting line of 1s and 0s would be more than five times as tall as Mount Everest. If you wrote out one gigabyte by hand, it would circumnavigate the globe at the equator. If you wrote out one terabyte by hand, it would extend to Saturn and back twenty-five times. If you wrote out one petabyte by hand, you could make a round trip to the *Voyager 1* probe, the most distant man-made object in the universe. If you wrote out one exabyte by hand, you would reach the star Alpha Centauri. If you wrote out all five zettabytes that humans produce each year by hand, you would reach the galactic core of the Milky Way. If instead of sending e-mails and streaming movies, you used your five zettabytes as an ancient shepherd might have—to count sheep—you could easily count a flock that filled the entire universe, leaving no empty space at all.

This is why people call these sorts of records big data. And today's big data is just the tip of the iceberg. The total data footprint of *Homo sapiens* is doubling every two years, as data storage technology improves, bandwidth increases, and our lives gradually migrate onto the Internet. Big data just gets bigger and bigger and bigger.

THE DIGITAL LENS

Arguably the most crucial difference between the cultural records of today and those of years gone by is that today's big data exists in digital form. Like an optic lens, which makes it possible to reliably transform and manipulate light, digital media make it possible to reliably transform and manipulate information. Given enough digital records and enough computing power, a new vantage point on human culture becomes possible, one that has the potential to make awe-inspiring contributions to how we understand the world and our place in it.

Consider the following question: Which would help you more if your quest was to learn about contemporary human society—unfettered access to a leading university's department of sociology, packed with experts on how societies function, or unfettered access to Facebook, a company whose goal is to help mediate human social relationships online?

On the one hand, the members of the sociology faculty benefit from brilliant insights culled from many lifetimes dedicated to learning and study. On the other hand, Facebook is part of the day-to-day social lives of a billion people. It knows where they live and work, where they play and with whom, what they like, when they get sick, and what they talk about with their friends. So the answer to our question may very well be Facebook. And if it isn't—yet—then what about a world twenty years down the line, when Facebook or some other site like it stores ten thousand times as much information, about every single person on the planet?

These kinds of ruminations are starting to cause scientists and even scholars of the humanities to do something unfamiliar: to step out of the ivory tower and strike up collaborations with major companies. Despite their radical differences in outlook and inspiration, these strange bedfellows are conducting the types of studies that their predecessors could hardly have imagined, using datasets whose sheer magnitude has no precedent in the history of human scholarship.

Jon Levin, an economist at Stanford, teamed up with eBay to examine how prices are established in real-world markets. Levin exploited the fact that eBay vendors often perform miniature experiments in order to decide what to charge for their goods. By studying hundreds of thousands of such pricing experiments at once, Levin and his co-workers shed a great deal of light on the theory of prices, a well-developed but largely theoretical subfield of economics. Levin showed that the existing literature was often right—but that it sometimes made significant errors. His work was extremely influential. It even helped him win a John Bates Clark Medal—the highest award given to an economist under forty and one that often presages the Nobel Prize.

A research group led by UC San Diego's James Fowler partnered with Facebook to perform an experiment on sixty-one million Facebook members. The experiment showed that a person was much more likely to register to vote after being informed that a close friend had registered. The closer the friend, the greater the influence. Aside from its fascinating results, this experiment—which was featured on the cover of the prestigious scientific journal *Nature*—ended up increasing voter turnout in 2010 by more than three hundred thousand people. That's enough votes to swing an election.

Albert-László Barabási, a physicist at Northeastern, worked with several large phone companies to track the movements of millions of people by analyzing the digital trail left behind by their cell phones. The result was a novel mathematical analysis of ordinary human movement, executed at the scale of whole cities. Barabási and his team got so good at analyzing movement histories that, occasionally, they could even predict where someone was going to go next.

Inside Google, a team led by software engineer Jeremy Ginsberg observed that people are much more likely to search for influenza symptoms, complications, and remedies during an epidemic. They made use of this rather unsurprising fact to do something deeply important: to create a system that looks at what people in a particular region are Googling, in real time, and identifies emerging flu epidemics. Their early warning system was able to identify new epidemics much faster than the U.S. Centers for Disease Control could, despite the fact that the CDC maintains a vast and costly infrastructure for exactly this purpose.

Raj Chetty, an economist at Harvard, reached out to the Internal Revenue Service. He persuaded the IRS to share information about millions of students who had gone to school in a particular urban district. He and his collaborators then combined this information with a second database, from the school district itself, which recorded classroom assignments. Thus, Chetty's team knew which students had studied with which teachers. Putting it all together, the team was able to execute a breathtaking series of studies on the long-term impact of having a good teacher, as well as a range of other policy interventions. They found that a good teacher can have a discernible influence on students' likelihood of going to college, on their income for many years after graduation, and even on their likelihood of ending up in a good neighborhood later in life. The team then used its findings to help improve measures of teacher effectiveness. In 2013, Chetty, too, won the John Bates Clark Medal.

And over at the incendiary FiveThirtyEight blog, a former baseball analyst named Nate Silver has been exploring whether a big data approach might be used to predict the winners of national elections. Silver collected data from a vast number of presidential polls, drawn from Gallup, Rasmussen, RAND, Mellman, CNN, and many others. Using this data, he correctly predicted that Obama would win the 2008 election, and accurately forecast the winner of the Electoral College in forty-nine states and the District of Columbia. The only state he got wrong was Indiana. That doesn't leave much room for improvement, but the next time around, improve he did. On the morning of Election Day 2012, Silver announced that Obama had a 90.9 percent chance of beating Romney, and correctly predicted the winner of the District of Columbia and of every single state—Indiana, too.

The list goes on and on. Using big data, the researchers of today are doing experiments that their forebears could not have dreamed of.

THE LIBRARY OF EVERYTHING

This book is the story of one of those experiments.

The object of our experiment was not a person or a frog or a molecule or an atom. Instead, the object of our experiment was one of the most fascinating datasets in the history of history: a digital library whose stated goal is to encompass every book ever written.

Where did this remarkable library come from?

In 1996, two Stanford computer science graduate students were working on a now-defunct effort known as the Stanford Digital Library Technologies Project. The goal was to envision the library of the future, a library that would integrate the world of books with the World Wide Web. They worked on a tool for enabling users to navigate through library collections, jumping from book to book in cyberspace. But this was not something that could be implemented in practice at the time, because relatively few books were available in digital form. So the pair took their ideas and techniques for navigating from one text to another, followed the big data trail to the World Wide Web, and turned their work into a little search engine. They called it Google.

By 2004, Google's self-appointed mission to "organize the world's information" was going pretty well, leaving founder Larry Page with some free time to get back to his first love, libraries. Frustratingly, it was still the case that only a few books were available in digital form. But something had changed in the intervening years: Page was now a billionaire. So he decided that Google would get into the business of scanning and digitizing books. And while his company was at it, Page thought, Google might as well do all of them.

Ambitious? No doubt. But Google has been pulling it off. Nine years after publicly announcing the project, Google has digitized more than 30 million books. That's about one in every four books ever published. Its collection is bigger than that of Harvard (17 million volumes), Stanford (9 million), Oxford's Bodleian (11 million), or any other university library. It has more books than the National Library of Russia (15 million), the National Library of China (26 million), and the Deutsche Nationalbibliothek (25 million). As of this writing, the only library with more books is the U.S.

Library of Congress (33 million). By the time you read this sentence, Google may have passed them, too.

LONG DATA

When the Google Books project was getting started, we, along with everyone else, read about it in the news. But it wasn't until two years later, in 2006, that the impact of Google's undertaking really sank in. At the time, we were finalizing a paper on the history of English grammar. For our paper, we had manually done some small-scale digitization of Old English grammar textbooks.

The books most relevant to our research were buried in the bowels of Harvard's Widener Library. Here's how to find them. First, go to floor 2 of the East Wing. Walk past the Roosevelt Collection and the Amerindian languages section; you'll see an aisle with call numbers 8900 and up. Our books were on the second shelf from the top. For years, as our research progressed, we made frequent trips to this shelf. We were the only people who had taken those books out in years, and sometimes in decades. No one cared much about our shelf but us.

One day, we realized that a book we had been using regularly for our study was now available on the Web, as part of the Google Books project. Curious, we started searching for other books on our shelf. They were there too. Not because the Google corporation cared about English grammar in the Middle Ages. Nearly every book that we checked, no matter what shelf it was on, now had a digital counterpart. In the time that it took us to examine a handful of books, Google had digitized a handful of buildings.

Google's books-by-the-building represented a completely new type of big data, and it had the potential to transform the way that people look at the past. Most big data is big but short: recent records produced from recent events. This is because the creation of the underlying data was catalyzed by the Internet, a relatively recent innovation. Our goal was to study the kinds of cultural changes that can span long time periods, as generation after generation of people lives and dies. When it comes to exploring changes on historical time scales, short data, no matter how big, isn't very useful.

Google Books is as big a dataset as almost any in our age of digital media. But much of what Google is digitizing isn't contemporary: Unlike e-mails, RSS feeds, and superpokes, the book record goes back for centuries. So Google Books isn't just *big* data, it's *long* data.

Since they contain such long data, digitized books aren't limited to painting a picture of contemporary humanity, as most big datasets are. Books can also offer a portrait of how our civilization has changed over fairly long periods of time—longer than the length of a human life, longer even than the lifetimes of whole nations.

Books are a fascinating dataset for other reasons, too. They cover an extraordinary range of topics and reflect a wide range of perspectives. Exploring a large collection of books can be thought of as surveying a large number of people, many of whom happen to be dead. In the fields of history and literature, the books of a particular time and place are among the most important sources of information about that time and that place.

This suggested to us that, by examining Google's books through a digital lens, it would be possible to build a scope to study human history. No matter how long it took us, we knew we had to get our hands on that data.

MO' DATA, MO' PROBLEMS

Big data creates new opportunities to understand the world around us, but it also creates new scientific challenges.

One major challenge is that big data is structured very differently from the kinds of data that scientists typically encounter. Scientists prefer to answer carefully constructed questions using elegant experiments that produce consistently accurate results. But big data is messy data. The typical big dataset is a miscellany of facts and measurements, collected for no scientific purpose, using an ad hoc procedure. It is riddled with errors, and marred by numerous, frustrating gaps: missing pieces of information that any reasonable scientist would want to know. These errors and omissions are often inconsistent, even within what is thought of as a single dataset. That's because big datasets are frequently created by aggregating a vast number of smaller datasets. Invariably, some of these component datasets are more reliable than others, and each one is subject to its own idiosyncrasies. Facebook's social network is a good example. Friending someone means different things in different parts of the Facebook network. Some people friend liberally. Others are much cagier. Some friend co-workers, but others don't. Part of the job of working with big data is to come to know your data so intimately that you can reverse engineer these quirks. But how intimate can you possibly be with a petabyte?

A second major challenge is that big data doesn't fit too well into what we typically think of as the scientific method. Scientists like to confirm specific hypotheses, and to gradually assemble what they've learned into causal stories and eventually mathematical theories. Blunder about in any reasonably interesting big dataset and you will inevitably make discoveries—say, a correlation between rates of high-seas piracy and atmospheric temperature. This kind of exploratory research is sometimes called "hypothesis free," since you never know, going in, what you'll find. But big data is much less incisive when it comes time to explain these correlations in terms of cause and effect. Do pirates bring about global warming? Does hot weather make more people take up high-seas piracy? And if the two are unrelated, then why are they both increasing in recent years? Big data often leaves us guessing.

As we continue to stockpile unexplained and underexplained patterns, some have argued that correlation is threatening to unseat causation as the bedrock of scientific storytelling. Or even that the emergence of big data will lead to the end of theory. But that view is a little hard to swallow. Among the greatest triumphs of modern science are theories, like Einstein's general relativity or Darwin's evolution by natural selection, that explain the cause of a complex phenomenon in terms of a small set of first principles. If we stop striving for such theories, we risk losing sight of what science has always been about. What does it mean when we can make millions of discoveries, but can't explain a single one? It doesn't mean that we should give up on explaining things. It just means that we have our work cut out for us.

A final major challenge is the change in where the data lives. As scientists, we are used to getting data by experimenting in our laboratories or going out into the natural world to write down our observations. Getting data is, to some extent, within the scientist's control. But in the world of big data, major corporations, and even governments, are often the gatekeepers of the most powerful datasets. And they, their citizens, and their customers care a great deal about how the data is used. Very few people want the IRS to share their tax returns with budding scholars, however well-intentioned those scholars might be. Vendors on eBay don't want a complete record of their transactions to become public information or to be made available to random grad students. Search engine logs and e-mails are entitled to privacy and confidentiality. Authors of books and blogs are protected by copyright. And companies have strong proprietary interests in the data they control. They may analyze their data with a view toward generating more ad revenue, but they are loath to share the heart of their competitive advantage with outsiders, and especially scholars and scientists who are unlikely to contribute to their bottom line.

For all these reasons, some of the most powerful resources in the history of human self-knowledge are going largely unused. Despite the fact that the study of social networks is many decades old, almost no public work has been done on the full social network of Facebook, because the company has little incentive to share it. Despite the fact that the theory of economic markets is centuries old, the detailed transactions of most major online markets remain largely inaccessible to economists. (Levin's eBay study was the exception, not the rule.) And despite the fact that humans have spent millennia striving to map the world, the images produced by companies like DigitalGlobe, which has created fifty-centimeter-resolution satellite images of the entire surface of the Earth, have never been systematically explored. When you think about it, these gaps in our usually insatiable human desire to learn and explore are shocking. This would be as if astronomers spent many lifetimes trying to study the distant stars, but for legal reasons were never permitted to gaze at the sun.

Still, just knowing that the sun is there can make the desire to stare at it irresistible. And so today, all over the world, a strange mating dance is taking place. Scholars and scientists approach engineers, product managers, and even high-level executives about getting access to their companies' data. Sometimes the initial conversation goes well. They go out for coffee. One thing leads to another, and a year later, a brand-new person enters the picture. Unfortunately, this person is usually a lawyer.

As we worked to analyze Google's library of everything, we had to find ways to deal with each of these challenges. Because the obstacles posed by digital books are not unique; they are merely a microcosm of the state of big data today.

CULTUROMICS

This book is about our seven-year effort to quantify historical change. The result is a new kind of scope and a strange, fascinating, and addictive approach to language, culture, and history that we call culturomics.

We'll describe all sorts of observations that can be made using a culturomic approach. We'll talk about what our ngram data has revealed about how English grammar changes, how dictionaries make mistakes, how people get famous, how governments suppress ideas, how societies learn and forget, and how—in little ways—our culture can appear to behave deterministically, making it possible to predict aspects of our collective future.

And of course, we'll introduce you to our new scope: a tool we created with Google, called—for reasons that will become apparent in chapter 3—the Ngram Viewer. Released in 2010, the Ngram Viewer charts the frequency of words and ideas over time. This scope—and the massive computation that led to its creation—is the robot historian of our opening vignette. You can try it yourself, right now, at http://books.google.com/ngrams. Ours is a hardworking robot, used by millions of people, of all ages, all over the world, at all hours of day or night, all hoping to understand history in a new way: by charting the uncharted.

In short, this book is about history as it is told by the robots, about what the human past looks like when viewed through a digital lens. And though today the Ngram Viewer might be seen as odd or exceptional, the digital lens is flourishing, much in the same way that the optical lens did centuries ago. Powered by our burgeoning digital footprint, new scopes are popping up every day, exposing once-hidden aspects of history, geography, epidemiology, sociology, linguistics, anthropology, and even biology and physics. The world is changing. The way we look at the world is changing. And the way we look at those changes . . . well, that's changing too.

How many words is a picture worth?

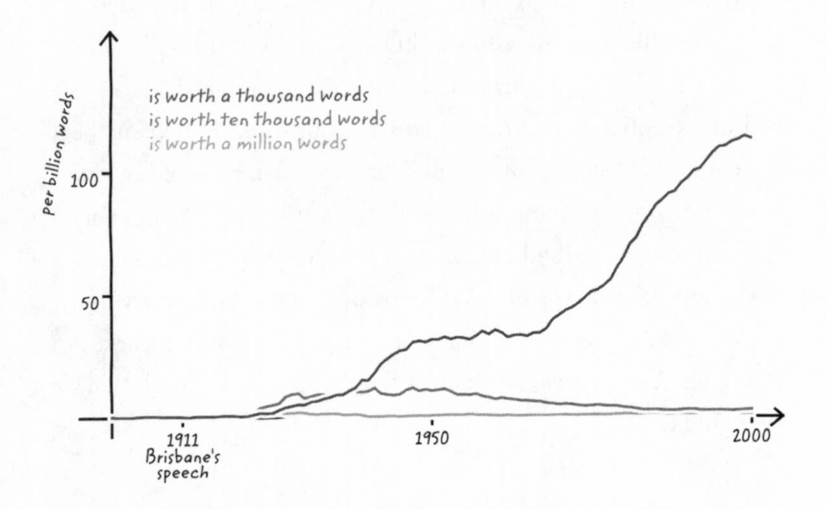

In 1911, the American newspaper editor Arthur Brisbane famously told a group of marketers that a picture is "worth a thousand words." Or he famously proposed that it's worth "ten thousand words." Or was it "a million words"? In any case, within decades, the expression had swept the country and—probably to Brisbane's chagrin—was now being billed as a Japanese proverb. (His listeners were in marketing, after all.)

What did Brisbane actually say? Alas, our new scope isn't likely to record the first instance of this expression. There's a Japanese proverb for that, too:

Compared to all speech,
Grasshopper, Google's scanned books
are but a haiku

Still, the scope can help us see how Brisbane's principle of iconic economics took shape.

It turns out that the *thousand words*, *ten thousand words*, and *million words* variants emerged shortly after Brisbane's (possibly) fateful remarks. All three forms competed for the next two decades. *Ten thousand* jumped to an early lead. But then came the '30s: Did *ten thousand* and *million* seem exorbitant to Depression-era ears? Whatever the cause, those years saw "a picture is worth a thousand words" begin the slow ascent that left its competition in the dust.

2

G. K. ZIPF AND THE FOSSIL HUNTERS

beautiful beautiful beautiful beautiful beautiful beautiful beautiful beautiful beautiful beautiful beautiful beautiful beautiful beautiful beautiful beautiful beautiful, beautiful, beautiful, beautiful, beautiful, beautiful, beautiful, beautiful," beautiful. beautiful. beautiful." beautiful . . . beautiful . . .

—*Legendary, Lexical, Loquacious Love*

In 1996, the concept artist Karen Reimer published the book *Legendary, Lexical, Loquacious Love.* Here is how she wrote it: She took the full text of a romance novel and alphabetized it. If a word appeared multiple times in the novel, it appears multiple times in her book.

The book has no syntax and no sentences. It is a 345-page-long list of words in alphabetical order. It does not look or read like a novel. In fact, when you read it, it appears to be complete nonsense.

We rarely read romance novels, but Reimer's work is an exception. An absolute page-turner, it fascinated us from cover to cover, from the dramatic beginning:

Chapter One

A

A A A A A A A A A A A A A A A A A A A A A A A A A A A A A
A A A A A A A A A A A

And all the way through the surprising finish:

Chapter Twenty-Five

Z

zealous

Twenty-five chapters, not twenty-six: There is no chapter for X, as the novel contained no words beginning with the letter X. Romance novels may be XXX-rated, but they contain very few actual X-words.

Even though it's just a single book, *Legendary, Lexical, Loquacious Love* gives really suggestive insights into the entire romance genre. For instance, it's clear that this is a book for her—the word *her* occupies almost a full eight pages (130–38). *His*? Two and a half (141–44). There's half a page of *eyes* and a third of a page of *breasts*, but only a single line about *buttocks*. Occasionally, the book is downright racy—there are three *climaxes* on page 62 alone. You go, girl! (Or guy; there's no way for us to know.)

Sometimes the book dwells too long on the superficial. For instance, *beautiful* appears twenty-nine times. *Intelligent*? Only once. But at other times, one gets a whiff of the original book's plot, such as a bone-chilling passage on page 187: "Murderers murderers, murdering murdering murdering murdering murdering murdering murdering, murderous murderous. murders murders, murky murmur murmured."

Over the years, we've turned to this book again and again, finding interesting new nuggets each time.

This is a bit odd. You would think that by alphabetizing a romance novel, and thereby obliterating its meaning, Reimer would also eliminate everything that made the novel interesting. And that's true, to an extent. But in the process, Reimer's alphabetical transmutation reveals a world that was once invisible: word frequencies, the lexical atoms from which the novel was composed. Those frequencies—and the stories they tell—are what make her work such an engaging read.

PROBLEM CHILD

When we met in 2005, big data was not yet a thing. The thought of reading millions of books in a split second hadn't entered our minds. We were just young graduate students looking to ply our trade on the most interesting questions we could find.

To find a fascinating question, it's helpful to have a fascinating environment. We met at Harvard's Program for Evolutionary Dynamics, a haven of creativity and science founded by the charismatic mathematician and biologist Martin Nowak. The PED (Program for Evolutionary Dynamics? Program for rEvolutionary Dynamics? Party Every Day?) is a place where mathematicians, linguists, cancer researchers, religious scholars, psychologists, and physicists congregate, thinking about new ways to look at the world. Nowak encouraged us to tackle the problems we found most interesting, regardless of where they might be found.

What makes a problem fascinating? No one really agrees. It seemed to us that a fascinating question was something that a young child might ask, that no one knew how to answer, and for which a few person-years of scientific exploration—the kind of effort we could muster ourselves—might result in meaningful progress. Children are a great source of ideas for scientists, because the questions they ask, though superficially simple and easy to understand, are so often profound. Questions like "Where does the sun go at night?" and "Why is the sky blue?" naturally lead the curious mind right into the heart of astronomy and physics. Questions like "Could a tree ever grow to be as tall as a mountain?" or "If we were really, really careful to avoid accidents, would we live forever?" turn on some of the most urgent issues in modern biology. "Why do I have to go to sleep?"—a tired cliché—still keeps neuroscientists up at night.

But of all these questions, one in particular caught our eye. "Why do we say *drove* and not *drived*?"

This question intrigued us because it was a simple example of a very profound concern about mankind. Why, as a culture, do we use certain words and not others? Why do we have certain ideas and not others? Why do we obey certain rules and not others?

Faced with a question like this, there are two possible approaches. One is to focus on the present circumstances that lead to a certain thing being a certain way. For instance: "Beloved child, you say *drove* because everyone else says *drove* and because, if you were to say *drived*, the neighbors would think that we, your parents, didn't bother to teach you proper English." This is a fine answer, which raises complex issues about the nature of social norms, issues that philosophers have been grappling with for centuries. But sometimes it can be more illuminating for a scientist to take the long view.

Surely the most impressive example of the long view in the history of the sciences lies in the work of Charles Darwin. More than 150 years ago, Darwin took a boat trip and encountered all sorts of creatures. He began to wonder about some birds that he saw in the Galápagos: Why are the beaks of those finches the way they are? More generally, why are all organisms the way they are?

What Darwin did next was extremely insightful. Instead of focusing entirely on the present, he took the long view. Darwin asked himself, How did things come to be this way over time? If we want to understand the world as it is now, Darwin reasoned, we must understand the process of change that brought about our present conditions. That process of change—Darwin's seminal discovery—is the combination of reproduction, mutation, and natural selection that together explain the remarkable diversity of the living world. In other words, the theory of evolution.

If you take the long view, the question of why we say *drove* and not *drived* becomes a scientific quest for the forces that shape the evolution of human culture. For a long time, we had no idea how to even begin to uncover those forces. All we had was a childlike question.

DINOSAUR HUNTERS

As scientists, we need to be able to collect data: cold, hard facts and precise measurements. We need to be able to frame unambiguous hypotheses, and then try to falsify them using definitive experiments and decisive analyses. From that standpoint, culture—

hard to define, harder still to measure—can be a tough nut to crack. This is what makes scientific work in fields like anthropology such an immense challenge, and is part of why, in 2010, the American Anthropological Association made the controversial decision to remove the word *science* from its statement of purpose. (The word has since been restored.)

We decided to start with a narrow aspect of culture that is much easier to define and to measure: language. Language is a great microcosm of the study of culture as a whole. It is the primary vehicle by which human culture is communicated. It changes, as is apparent to anyone who's ever wound up in the audience of one of Shakespeare's plays. Finally, language is often written down, and in that form furnishes a convenient dataset for scientific analysis. After all, written language is one of the earliest ancestors of big data.

So how should we go about exploring the evolution of language? In biology, there is no better way to understand the broad patterns of evolution than by looking at fossils. But finding fossils is hard. It requires a combination of careful planning and good strategy. If we hope to make progress finding fossils, we would do well to learn from Nathan Myhrvold, perhaps the greatest dinosaur hunter of his generation. (A man of many talents, he also founded Microsoft Research and wrote the book on modernist cuisine.) It's not that Myhrvold is luckier than everyone else, and that every whitish rock he haplessly blunders across turns out to be a *T. rex* skull. Myhrvold and his team use detailed geological maps, satellite images, and their own painstaking analysis of *T. rex* ecology to decide where to explore, where the whitish rocks are likeliest to be fossils. As a result, they've found nine *T. rex* skeletons since 1999—when only eighteen such skeletons had been found in the ninety previous years. As Myhrvold puts it, "We have dominant *T. rex* market share."

Our ambition was to get dominant language-fossil market share. Just as dinosaur fossils tell us about biological evolution, linguistic fossils would help us understand how language evolves. But if we wanted to have a good chance of finding such fossils, we needed some kind of guiding principle to help us figure out where to dig. As it turns out, just such a compass had been created eighty years ago, by a man who, like us, really liked to count.

1937: A DATA ODYSSEY

George Kingsley Zipf was at Harvard in the 1930s and 1940s, and was chair of the German literature department. He had a mix of skills that is rather rare: a prominent humanist, but with a very quantitative bent.

Being a man of letters, Zipf spent a lot of time thinking about words. It was rather obvious to Zipf that all words are not created equal. The word *the* is used all the time, but we rarely hear the word *quiescence*. Zipf found this imbalance puzzling and wanted to understand what was going on.

One way to think about Zipf's question is as follows: Imagine that the English language were a nation, and each word a citizen. And imagine that the height of each word-person were proportional to the frequency of that word's use—*the* would be a giant word, but *quiescence* would be tiny. What would it be like to live among such oddly sized people? That's the kind of childlike question Zipf found fascinating.

To picture what this world looks like, Zipf needed to take a census of all the words and count how many times each one was used. Today, this kind of thing is computationally trivial (a one-line command). That's why the concept art book *Legendary, Lexical, Loquacious Love* didn't take decades to write. But back in 1937, nothing was computationally trivial. Modern computers didn't exist. The word *computer* meant a researcher whose job was to perform arithmetic calculations.

If he was going to count words, Zipf would have to do it the old-fashioned way, by recording every instance of every word, one by one, by hand. Of course, that would be soul-crushingly boring.

He must have been pretty ecstatic when he came across the work of Miles L. Hanley. Hanley, who was a big fan of *Ulysses*, had published a painstaking and heroic work that he had given the rather boring title *Word Index to James Joyce's* Ulysses. This book, a type of scholarly work known as a concordance, was meant to allow fellow *Ulysses* scholars and enthusiasts to find every instance of any word in the book. To Zipf, no book could have been more exciting. In order to get at his original problem, all Zipf had to do was take Hanley's index and count how long each of the entries was. Much, much easier.

Note that Zipf understood, well ahead of his time, what scientists and humanists today are just beginning to learn: how to follow the data. Zipf skillfully reframed the questions he cared about in light of the kind of data available to him. Instead of tackling the impossible problem of counting all words, he settled for the tractable problem of counting words in *Ulysses*. If he were alive today, he would have been at Google's door the moment the company announced the book digitization project.

Equipped with Hanley's index, Zipf ranked the words in *Ulysses* by their frequency. The top spot is taken by *the*, used

14,877 times—one out of every eighteen words. The tenth most frequent word is *I*, with 2,653 appearances. *Say*, which appeared 265 times, comes in at one hundredth place. *Step*, which occurs 26 times, appears in the thousandth spot on Zipf's ranked list. To be tied for the ten thousandth position, like the word *indisputable*, a word needed only appear twice.

As he looked over his ranked list, Zipf noticed something funny. There was an inverse relationship between the rank of a word and its frequency of use. If a word's numerical rank was ten times as high—five hundredth place instead of fiftieth—then it was ten times as rare. So *his*, ranked eighth with 3,326 mentions, is ten times more frequent than *eyes*, ranked eightieth, which appears 330 times. An equivalent way of thinking about this is to say that there are far more rare words than you might expect. In *Ulysses*, only ten words are used more than 2,653 times. But there are a hundred words used more than 265 times, and a thousand words used more than 26 times, and so on and so forth.

And, as Zipf soon discovered, this wasn't just a feature of words in Joyce's *Ulysses*. The same regularity appeared in words taken from newspapers, texts written in Chinese and Latin, and pretty much everywhere else he looked. Called Zipf's law today, the discovery turned out to be a universal organizing principle of all known languages.

THE WORLD ACCORDING TO ZIPF

Before Zipf, scientists thought that most things you could measure behaved like human height.

Human height doesn't vary terribly much. Ninety percent of

the adults in the United States are between five feet and six-foot-one. Sure, some extremely tall basketball players are seven and a half feet, and the world's smallest adult is just under two feet tall. But both cases are very, very rare. And even when you consider these extremes, the tallest people are only four to five times as tall as the shortest. Mathematicians have a special word for this kind of distribution, where the values are so tightly clustered around an average. They call this commonly observed distribution "normal." Before Zipf, people thought we lived in a normal world, where things were all normal.

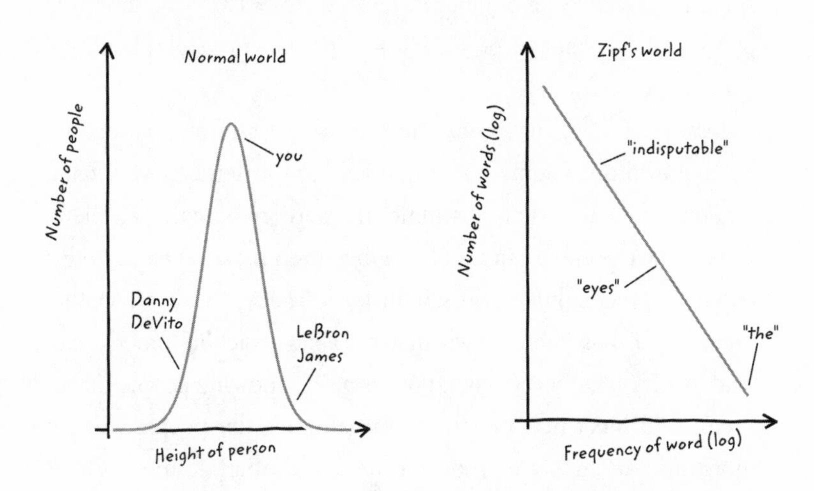

But as we've seen, the world of words is far from normal, with a distribution of sizes that obeys a very specific, and seemingly strange, mathematical pattern. Today, scientists call these behaviors power laws. Surprisingly, once Zipf found his first power law in language, he started to find them everywhere.

For instance, Zipf found that both wealth and income exhibit power laws. If your height were proportional to your bank account, and the average American household were five-foot-seven, then Bill Gates would be taller than the moon. The lengths of articles in the *Encyclopædia Britannica* also obey a power law, as do newspaper circulation rates. Scientists following up on Zipf's work found thousands of other examples: the size of cities, the frequency of particular last names, the bloodiness of wars, how long people clap after a performance, the popularity of people on Facebook and Twitter, the amount of food consumed by animals, the traffic at Web sites, the abundance of proteins in our cells, the abundance of cells in our bodies, the abundance of species in our ecosystems, and the size of holes in Swiss cheese. Even the length of power outages obeys a power law—or perhaps we should call it a "lack of power" law.

Although Zipf's work was transformative, the reasons behind his ubiquitous law remain mysterious. Zipf himself believed that it emerged because such distributions were maximally efficient. Others have pointed out that being big often makes it easier to get bigger, a process known to scientists as "rich get richer." Mathematically, it has been shown that a "rich get richer" process can lead to all sorts of power laws. For instance, knowing people makes it easier to meet new people, so initially popular people will get more and more popular over time, in Zipfian fashion. Cities that are already large might be appealing to someone considering a move, leading to power laws of city size. Here's yet another account: It can be shown that monkeys typing on a computer at random would produce "words" (characters separated by a space) whose abundance exhibits a power law.

There are often multiple competing explanations for the cause of any particular power-law distribution. Alas, this overabundance of explanations probably reflects the fact that scientists don't really know what's going on.

Still, whatever their cause, power laws aptly describe a stunning range of natural and social phenomena. Zipf, a professor of German, aided by Hanley's uncommon enthusiasm for the novel *Ulysses*, set off a revolution whose consequences transformed much of quantitative social science and whose tentacles have reached biology, physics, and even mathematics. Zipf is the new normal.

TOO ZIPF OR NOT TOO ZIPF

Zipf's law was just the touchstone we needed to go hunting for remnants of language evolution. Virtually everything in language obeys Zipf's law: nouns, verbs, adjectives, adverbs that start with *m*, words for professions, words that rhyme with *rhyme*, and so on and so forth. So, when you run into something that does *not* behave according to Zipf's universal principle, something really fishy is going on. Like a whitish rock that seems out of place at a particularly promising expedition site, a phenomenon in language that doesn't obey a power law just might turn out to be a fossil of our language's evolution.

That's where the childlike question that had so captivated us comes in: "Why do we say *drove* and not *drived*?"

Drove is a member of a class of English words called the irregular verbs. Irregular verbs are strange. If irregular verbs obeyed Zipf's law, the way almost all other classes of words do, you would expect most of them to be rare. Instead, nearly all irregular verbs are very frequent. Although only about 3 percent of verbs are irregular, the ten most frequent verbs are all irregular. To put it simply, irregular verbs are a clear and dramatic exception to Zipf's law. They were exactly what we had been looking for, as though the position of the *T. rex* skeleton had been conveniently marked by a statistical headstone.

Who were these so-called irregular verbs, what had they done to Zipf, and what did that mean about the evolution of language?

THE FEW, THE PROUD, THE STRONG

English verb conjugation is, at first glance, a walk in the park. To form the past tense of an English verb, all you have to do is to add *-ed*: *jump* becomes *jumped*. Hundreds of thousands of verbs obey this simple rule. When new verbs enter the language, they obey this rule by default. I may have never heard of *flamboozing* before, but I know that if you chose to *flambooze* yesterday, then yesterday you *flamboozed*.

Except—much to the chagrin of English learners—for the pesky irregular verbs. Verbs like *to know*. Even before you read this sentence, you probably *knew* that we don't say *knowed*. About three hundred in all, the irregular verbs—sometimes called strong verbs by linguists—include the ten most frequent verbs in the English language: *be/was, have/had, do/did, say/said, go/went, get/got, make/made, know/knew, see/saw, think/thought*. They are so frequent that, when you use a verb, there is a 50 percent chance that it will be irregular.

Where did the irregulars come from? It's a long story. Sometime between six thousand and twelve thousand years ago, a language known to modern scholars as Proto-Indo-European was spoken. An astonishing array of modern languages, including English, French, Spanish, Italian, German, Greek, Czech, Persian, Sanskrit, Urdu, Hindi, and hundreds of others, descend from Proto-Indo-European. Proto-Indo-European had a system, known to scholars as the ablaut, that transformed a word into a related one by changing its vowels according to fixed rules. In English, the ablaut can still be seen in the form of subtle patterns among the irregular verbs.

Here is an example of one pattern: Today I *sing*, yesterday I *sang*, the song was *sung*. Similarly: Today I *ring*, yesterday I *rang*, the phone has *rung*. Here's another pattern: Today I *stick*, yesterday I *stuck*. Today I *dig*, yesterday I *dug*. When rules of conjugation die, they leave behind fossils. We call those fossils irregular verbs.

What sort of grammatical asteroid wiped out these ancient rules, leaving behind only the dry bones of the irregulars?

That asteroid was the so-called dental suffix, written as *-ed* in Modern English. The use of *-ed* to signify the past tense emerged in Proto-Germanic, a language spoken between 500 and 250 BCE in Scandinavia.

Proto-Germanic was the linguistic ancestor of all the modern Germanic languages, including English, German, Dutch, and many others. Because it was a descendant of Proto-Indo-European, Proto-Germanic inherited the old ablaut scheme for conjugating verbs. And this worked fine most of the time. But occasionally, new verbs entered the language, and some of these didn't quite fit any of the old ablaut patterns. So the speakers of Proto-Germanic invented something new, forming the past tense of these young, nonconformist verbs by adding that *-ed*. In Proto-Germanic, the regular verbs were the exception.

But not for long. Use of the dental suffix to mark the past tense was a tremendously successful invention, and it began to spread rapidly. Like any disruptive technology, the new rule started at the margins, serving funky-looking verbs that the ablaut could not. But once it had established this beachhead, it did not stop. Simple and memorable, the dental suffix began to attract additional adherents, as verbs that had always used the venerable ablaut patterns started making the switch.

Thus, by the time that the classic Old English text *Beowulf* was written, about 1,200 years ago, more than three-quarters of English verbs obeyed the new rule. With its strength eroded, the old ablaut was now on the run, the upstart *-ed* rule everywhere nipping at its heels. More and more irregular forms defected over the next thousand years. A millennium ago, I would have *holp* you. Just yesterday, though, I would have *helped* you.

This is a process that today's linguists, with the benefit of hindsight, call regularization. And it's still going on. Consider the verb *thrive*. About ninety years ago, a headline in the *New York Times* read "Gambling Halls *Throve* in Billy Busteed's Day." But in 2009, the *Times* ran an article in its Science section titled "Some Mollusks *Thrived* After a Mass Extinction." Unlike those lucky mollusks, *throve* was a victim of the mass extinction of the ablaut. There is no going back: Once they are regular, verbs almost never irregularize. For every *sneak* that *snuck* in, there are many more *flews* that *flied out*.

Like the three hundred Spartans at Thermopylae, the English irregular verbs—three hundred, strong—have been resolutely holding off a merciless assault on their kind that began in 500 BCE. It is a battle they have waged every day, in every city, in every town, along every street where English is spoken. They have been waging it for 2,500 years. They are not merely exceptions: They are survivors.

And the process that they survived was exactly the process that we intended to study: the evolution of language.

2005: ANOTHER DATA ODYSSEY

Why did certain irregular forms die out, while others managed to survive? Why didn't *throve* thrive on, and why didn't *drove* drive off?

Linguists already had some great ideas about why irregular verbs have such high frequencies. They reasoned that the less often we encounter an irregular verb, the harder it is to learn and the easier it is to forget. Because of this, rare irregular verbs, like *throve*, disappear more rapidly than the frequent ones, like *drove*. Over time, low-frequency irregulars drop out, and the irregulars as a whole become more frequent.

To us, this hypothesis was extremely exciting, as it suggested that irregular verbs are undergoing a process identical to evolution by natural selection. Why are the irregulars so frequent, when, in accordance with Zipf's law, every other lexical class is dominated by rare words? Because natural selection, in the form of the insatiable *-ed* rule, gives common irregulars an evolutionary advantage. The more frequent a verb is, the more fit it is to survive.

This was by far the tidiest account of natural selection operating on human culture that we had ever encountered. Zipf's compass had guided us to a fascinating problem: Would the linguists' hunch be borne out under careful scrutiny? If so, it would be a simple illustration that human culture can evolve by natural selection. Like Zipf, all we had to do now was find the data.

To aid us in our quest, we enlisted two extremely bright undergraduates at Harvard College, Joe Jackson and Tina Tang. In an ideal world, we hoped that Joe and Tina could read everything ever published in the English language and record every instance of an irregular verb that they encountered. But they told us that they were both planning on graduating in four years. (As doctoral students, the thought of graduating rarely crossed our minds.) We would need to improvise.

Fortunately, Joe and Tina had learned a great deal from the story of Zipf. They hit on an alternative approach. Instead of reading absolutely everything, why not just read all the textbooks on historical English grammar? Grammar texts of, say, Middle English, would surely discuss irregular verbs, would mention many of them, and would probably provide a partial listing somewhere. By going through the library and reading every single textbook dealing with the grammar of historical Englishes, we could probably get a pretty good picture of what was irregular and when. These grammar textbooks could do for us exactly what Hanley's *Ulysses* treatise did for Zipf.

Of course, this is easier said than done. Joe and Tina did many months of meticulous work, reading textbooks of Old English (the language of *Beowulf*, spoken circa 800 CE) and of Middle English (Chaucer's language, spoken around the twelfth century). They dug up 177 Old English irregular verbs, each of which they could track for more than a thousand years. With a millennium's worth of snapshots, we could finally see how the language was changing.

All 177 verbs started as irregulars in Old English. By the time of Middle English, four centuries later, only 145 of the irregular forms survived; the remaining 32 had regularized. By Modern English, only 98 of them remained irregular. The other 79 verbs are still in the language, but, like *melt*, they have changed form.

Yet there was a striking imbalance. Among the 12 most frequent verbs in our list, none had become regular—they had all resisted twelve centuries of pressure by the *-ed* rule. At the other end of the spectrum, the casualties were everywhere. Of the 12 least frequent verbs on our list, 11 had become regular, including verbs like *bide* and *wreak*. The only low-frequency irregular to survive is *slink*, a verb that aptly describes this quiet process of disappearance.

The data had spoken: Something akin to natural selection was influencing human culture, leaving its fingerprints among the verbs. Usage frequency was having an extraordinarily strong effect on verb survival, making the difference between the verbs that were *mourn/mourned* and the verbs that were *fit/fit* to survive.

SURVIVAL OF THE FIT

In biology, it is much easier to show that natural selection for a trait is happening than to measure the exact relationship between that trait and evolutionary fitness. (It's easy to tell that it's windy, but much harder to tell how strongly the wind is blowing.) Without estimates of fitness, all we know is what sort of changes evolution will favor; we have no idea how long it will take for those changes to come about.

The case of the irregular verbs, though, is not like the typical case of biological evolution. In biology, thousands or even millions of traits must be taken into account to compute the fitness of a single organism. For the irregulars, it was clear that a single trait—usage frequency—was by far the most significant factor in determining fitness. This simplified matters immensely. It meant that we might be able to reliably estimate how quickly the irregular forms of verbs would disappear.

But before we dive into that, let us remind you of the most famous disappearing act in all of science: the theory of radioactivity.

Radioactive materials are used in everything from power reactors to medical imaging systems to bombs. These materials are constantly in the process of disappearing, because, as time passes, atoms of a radioactive substance morph into stable, nonradioactive atoms. This decay releases energy, often in the form of radio waves. That's how radioactive substances got their name.

The most important property of a radioactive substance is its half-life. This is the period of time it takes, on average, for half of the atoms in a sample of the substance to decay. Suppose you have a substance whose half-life is one year. If you start with a billion atoms of that substance in a jar, then a year later, only half a billion atoms of the substance will be left—the other half billion will have decayed into something else. After two years, only one-quarter of a billion atoms (half of a half) will be left. After three years, an eighth. And so on.

As we examined the transformation of irregular verbs into regular verbs, we found that, once one took frequency into account, the process of regularization was mathematically indistinguishable from the decay of a radioactive atom. Moreover, if we knew the frequency of an irregular verb, we could use a formula to compute its half-life. This was remarkable, because for radioactive atoms, you have to measure the half-life experimentally; it's usually impossible to compute. In this respect, the mathematics of radioactivity applied even more neatly to irregular verbs than to radioactive atoms.

The formula was simple and beautiful: The half-life of a verb scales as the square root of its frequency. An irregular verb that is one hundred times less frequent will regularize ten times as fast.

For instance, verbs whose frequencies fall between one in one hundred and one in one thousand—verbs like *drink* or *speak*—have a half-life of roughly 5,400 years. This is comparable to the half-life of carbon-14 (5,715 years), the isotope that is most famously used in dating ancient relics.

THE ONCE AND FUTURE PAST

Once you've calculated the half-life of irregular verbs, it's possible to make predictions about their future. Based on the above analysis, we predicted that by the time one verb from the set *begin, break, bring, buy, choose, draw, drink, drive, eat, fall* regularizes, five verbs from the set *bid, dive, heave, shear, shed, slay, slit, sow, sting, stink* will have already regularized. And that if current trends hold up, only 83 of our 177 irregular verbs will still be irregular in the year 2500.

We were so excited about this that we summed our predictions up as a short story:

> He was a well-breeded man from the twenty-sixth century, so it really stinged when they said his grammar stunk. "Stinked," the time-traveler corrected.

If you're planning on doing some time travel anytime soon, you'd do well to memorize this instructive tale.

We could also anticipate the fate of particular verbs. After thousands of years together, which of today's irregular verbs is most likely to abandon its current conjugal partner in pursuit of a younger model? Paradoxically, the answer is *wed/wed,* least frequent of the modern irregular verbs. Already, *wed/wedded* are frequently spotted in public. Now is your last chance to be a *newly-wed*. The married couples of the future can only hope for *wedded* bliss.

And finally, we could answer the childlike question that had started us off on our journey.

"Why do we say *drove* and not *drived*?"

The reason we still say *drove*—whereas we've abandoned other irregular forms, like *thrive,* in droves—is that *drive* is far more frequent than *thrive*. In any given century, verbs like *throve* are about five times as likely to regularize as verbs like *drove*. Of course, *drove,* too, will eventually disappear, if English survives long enough. Our estimates suggest that we still have about 7,800 years before *drove* drives off into the sunset. Kids will keep on wondering about it for a long time to come.

JOHN HARVARD'S SHINY SHOE

In the center of Harvard Yard, there is a big statue commemorating the life of John Harvard. The bronze figure has a dull coloration, except for the left shoe, which always looks shiny. For some reason, taking a picture with your hand touching this shoe has become an item on the to-do list of every tourist who visits Harvard.

Why is John Harvard's shoe so shiny? Most people think that when the sculpture was originally created, the entire façade—including the footgear—was a dull bronze, and that gradual polishing by thousands of visiting hands first exposed the shoe's gleaming surface.

But bronze is a naturally shiny metal. When it was originally cast, more than a century ago, this sculpture—like any other bronze sculpture—was shiny, too. The absence of luster, a topmost layer known as the sculpture's patina, is a result of corrosion brought about by natural weathering, by restoration efforts, and even by the artist himself. The metal's true color survives only in that shoe, thanks to the frequent brush of thousands of passersby.

The irregular verbs are just like this. When you first encounter them, you wonder, How did these strange exceptions get here? But in fact, the irregular verbs obey the same patterns today that they obeyed many centuries ago. As the language around them changed, frequent contact protected the irregulars from corrosion. They are fossils of an evolutionary process that we are just beginning to understand. Today, we call all those other verbs regular. But regularity is not the default state of a language. A rule is the tombstone of a thousand exceptions.

LEXICON AND CONCORD

The *Word Index to James Joyce's* Ulysses was a triumph, reflecting years of perseverance and attention to detail. At the time it was published, in 1937, such indices were available for only the most important books, despite the fact that concordance writing has an extremely long and illustrious history. For instance, the oldest concordances of the Hebrew Bible, known as the Masorah, were penned more than a thousand years ago.

Things began to change in 1946. That year, a Jesuit priest named Father Roberto Busa had a powerful idea. Busa, a scholar of the prolific theologian Thomas Aquinas, wanted a concordance of Aquinas' work to help him with his studies. Computer technology was beginning its meteoric ascent, and Busa thought it might be possible to create a concordance in a new way, by feeding the raw text of a book into one of these new machines. He took his idea straight to IBM. The company heard him out and decided to support his efforts. It took thirty years and lots of IBM's help, but Busa's plan eventually worked: The monumental *Index Thomisticus* was completed in 1980. The world of scholarship was impressed, and like Hanley's *Index*, Busa's *Index* ultimately gave rise to a new field. Known today as the digital humanities, work in this area concerns itself with all the ways in which computers can be relevant to traditional humanistic enterprises like history and literature.

Despite the extraordinary influence of these indices, it is easy to think of them as a swan song. It was not long before the burgeoning power of modern computers meant that creating a concordance took only a single line of code, easy to write and instantaneous to run. By the time Reimer published the alphabetical experiment she called *Legendary, Lexical, Loquacious Love*—essentially a concordance, but with page references left out—the concordancing itself merited only a brief acknowledgment. Today, scholars rarely bother to make new concordances. There's no need, since a cheap laptop computer can search a long text for all instances of a word almost instantaneously. On the surface, the age of concordances has come to an end.

Yet if you pop the hood of modern technology, what you find underneath may surprise you. Today's world is kept humming by Internet search engines, the most powerful information-finding tools ever developed. What is a search engine? At its core, a search engine is a list of words and the pages on the Web on which those words appear. Hiding behind every little white search box is a massive digital concordance.

Concordances didn't die out after Busa. Instead, they took over the world.

TAKING ROSES APART TO COUNT THEIR PETALS

Zipf was a remarkable man whose work transformed numerous fields, most of them distant from his own expertise. From language to biology, urban planning to the physics of cheese, it's hard to be a scientist today without encountering Zipf's legacy. In our own work, Zipf provided the clue we needed to begin uncovering the secrets of language evolution.

What was it about this oddball scholar of German literature that made him, scientifically speaking, so prophetic?

George A. Miller, one of the founders of cognitive psychology, once gave the following take on Zipf, and we think it goes a long way toward answering that question. Miller said that Zipf was the kind of man who would "take roses apart to count their petals." On the surface, this doesn't sound terribly flattering. Was Zipf so obsessed with counting that he was unable to appreciate the beauty of a flower?

Certainly not. Zipf was a prominent scholar of literature,

someone who deeply grasped the beauty and power of the book, the flower of literary genius. What made Zipf different, though, was that he wasn't so transfixed by this beauty as to be blind to the other ways in which a flower could be appreciated. One of those ways happened to involve taking the flower apart.

Before Zipf, a book was something that was read, understood, and contemplated line by line and page by page. You took in the whole gestalt, like a rose in full bloom. Even Hanley, whose index had facilitated Zipf's journey, intended his work as an aid for traditional reading.

But embedded in Zipf's peculiar question was a radical new notion of what a book could be. The question reflected his marvelous intuition that an alternative form of reading was possible: analyzing little petals of text, stripped of their floral context, to look for evidence of mathematical design.

For the last century, scientists have been following the trail of this pioneering insight. By the time we had finished our analysis of verbs, we were proud to count ourselves among their number. But in truth, we were still too caught up with the particulars of the irregular verbs to really appreciate the power of Zipf's approach.

That would soon change. After all, Zipf had revealed breathtaking scientific horizons by picking apart a mere handful of flowers. Now, thanks to Google, entire libraries were being digitized, one after another. We wanted to try what Zipf had done. But we wanted all the flowers.

Burnt, baby, burnt

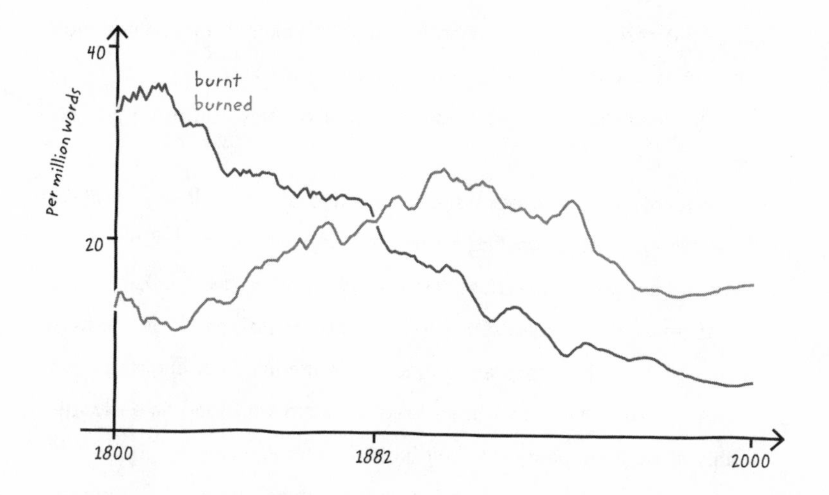

Studying English in his native land, a young Frenchman *learnt* that certain verbs were *spelt* differently in the past tense. These *spoilt* verbs *dwelt* in their own section of the textbook, singled out even among the irregulars. Although it was a real pain in the neck to learn them all, he soldiered through, memorizing the list of words whose past tense was formed by adding *-t* instead of *-ed*.

When he finally entered the United States, the student was brimming with confidence in his mastery of the language. But shortly after his arrival, while reading about the London Olympics in the press, he was surprised to see the following headline in the *Washington Post*: "Burned-Out Phelps Fizzles in the Water against Lochte." As every Frenchman is taught, the verb *burn* is irregular. Michael Phelps should have felt *burnt* out. Didn't these American papers have copy editors?

A few days later, he saw another distressing headline, this one in the *Los Angeles Times*: "Kobe Bryant Says He Learned a Lot from Phil Jackson." The student knew nothing about Phil Jackson, but was still shocked that Kobe had *learned* from Phil. If anything, he should have *learnt*.

Little by little, the student realized that when it came to this particular rule, all Americans were making the same mistakes. He knew that most Americans sounded ludicrous when they spoke French, but to judge from his textbooks, they were equally bad at their native tongue. He *smelt* a rat.

Fortunately, he had access to a new kind of scope. It soon *spilt* the beans: He had been wasting his time back in France. He felt *burnt*.

What happened? Because the verbs *burn/burnt*, *dwell/dwelt*, *learn/learnt*, *smell/smelt*, *spell/spelt*, *spill/spilt*, and *spoil/spoilt* all follow a similar pattern, they prop each other up in the minds of English speakers. As a result, they have been irregular for a very long time—longer than you would expect from their individual frequencies.

These verbs still appear as irregular in many textbooks. But in reality, the once-mighty alliance is coming apart. Two members, *spoil* and *learn*, regularized by 1800. Four more have regularized since then: *burn*, *smell*, *spell*, and *spill*.

The results suggest that this trend originated in the United States. But it has since spread to the United Kingdom, where each year, a population the size of Cambridge, England, adopts *burned* in lieu of *burnt*. Today, only *dwelt* still dwells among the irregulars.

In conclusion: The student was wrong to feel *burnt* by his English language courses. He should have felt *burned*.

3

ARMCHAIR LEXICOGRAPHEROLOGISTS

By 2007, our encounter with irregular verbs had convinced us that counting words made it possible to track certain kinds of cultural change over time. But tracking irregular verbs is easy, because they are so frequent. The word *went*, for instance, appears about once every five thousand words, or roughly every twenty pages. You see it repeatedly in every book you read. But as one ventures beyond the irregular verbs, trying to track words more generally, one soon runs into the dark side of Zipf's law. The words that are frequent (like *went*) are very few in number. The vast majority of words are exceedingly rare.

Suppose we were trying to track something a bit more challenging, like the abominable snowman known as the *Sasquatch*. The elusive *Sasquatch* appears in English texts approximately once in every ten million words, or roughly once every hundred books. Tracking down the *Sasquatch* is much, much harder than tracking the typical irregular verb.

Still, as cultural concepts go, the *Sasquatch* isn't very hard to find. The *Loch Ness monster* is more elusive—only one appearance every two hundred books. But if you want to really test your mettle as a lexical tracker of cryptic creatures, try finding a *Chupacabra*. The blood-drinking creature was first spotted in 1995 in Puerto Rico. Not much more is known. But we can tell you this: A *Chupacabra* is much rarer than a *Sasquatch*. There's a sighting just once in every 150 million words, or about 1,500 books. An extremely well-read person might see a *Chupacabra* once in his or her entire life. Here it is, one last time: *Chupacabra*. Savor this moment.

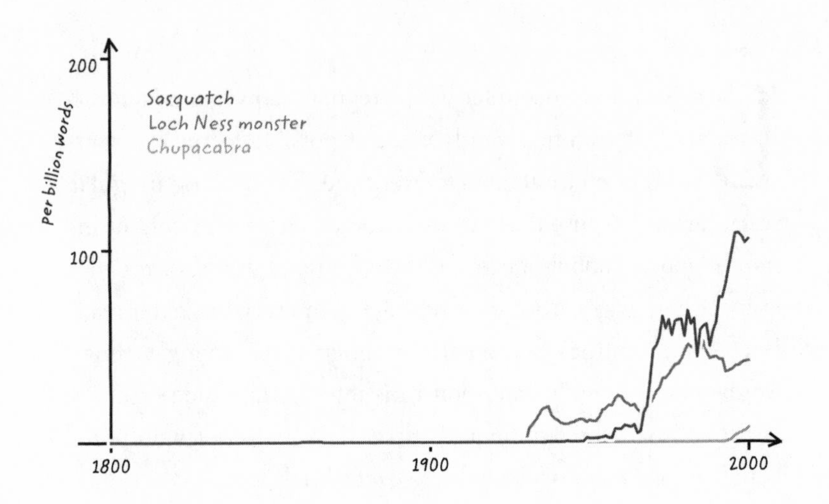

To track words like that one, we'd need millions of books at our disposal: big data. And there was only one place we could go to get it.

TWENTY-NINE-YEAR-OLD BILLIONAIRE PSYCHOLOGY

In 2002, Google was going great guns, and cofounder Larry Page had some free time. What to do? Google's mission is, after all, to "organize the world's information," and Page knew that there's a lot of information in books.

He began to wonder: How hard would it be to transform a physical, brick-and-mortar library into a digital one that could live in cyberspace? No one knew. So Page and Marissa Mayer (then a product manager at Google; as of 2013, the CEO of Yahoo!) decided to do an experiment, using a metronome to help them keep the pace as they turned the pages of a three-hundred-page book. It took forty minutes. At that rate, just flipping through the pages of a seven-million-volume library, such as that of Page's alma mater, the University of Michigan, would take about five hundred years. And of course, the University of Michigan has only a fraction of all books. Flipping through the pages of all the world's books—as you would need to do in order to digitally scan each page into a machine-readable form—would take millennia, even eons. It seemed impossible.

But of course, you're not thinking like a twenty-nine-year-old billionaire. To a giant of the Internet boom whose company would soon enter the Fortune 500, a person-eon is a commodity that you can buy.

So when the University of Michigan's president, Mary Sue Coleman, told Page that completely digitizing the university's books would take a thousand years, Page offered Google's services and suggested that the task could be completed in six.

And with that, Google began a project to digitize every single book ever written—to assemble a library of everything, and load it onto a computer hard drive.

PAGE'S PAGES

Before Google could go about acquiring and scanning all the books, the company needed a shopping list to help keep track of which books it needed to get and which it had already scanned. So Google collected book catalog information from hundreds of libraries and companies, and then merged these catalogs to create a list containing, as best Google could tell, an entry for every book ever written. (Or, more precisely, for every book that has survived into the present day. The books lost when the Library of Alexandria burned down, for instance, don't count in this total.) The resulting shopping list contained 130 million books.

Next, it needed to acquire and scan each book. In some cases, publishers sent copies straight from the presses. In this situation, Google would scan the books "destructively": Employees would cut off the binding and scan the pages in, one after another, at very high speed, storing the images in a digital format that could easily be viewed on a computer. For the rest of the books, the company reached out to libraries around the world, checking out shelves, sections, wings, and even whole buildings at a time. Like all library books, the volumes needed to be returned—even Google couldn't hope to afford all those late-book fees. So Google developed a nondestructive scanning system, too: A small army of page turners, following in Page and Mayer's footsteps, was hired to turn pages all day long while cameras snapped images of the text. In the last decade, this unstoppable scanning squadron has turned the page billions of times. Every once in a while, a telltale thumb appears in one of the images.

Finally, using a process called optical character recognition, in which a computer program finds and identifies the letters contained in an image, the digitized images are transformed into raw text. The result is a text file—akin to what you might produce when typing in a word processor—that contains the entire book.

In a major triumph for twenty-nine-year-old billionaire logic, Google's digitization efforts have been extraordinarily successful. Ten years after Page flipped pages with Mayer and nine years after he publicly announced the project, Google has digitized more than thirty million books.

Such a vast collection of text can only be analyzed by computer. If a human tried to read it, at the reasonable pace of two hundred words per minute, without interruptions for food or sleep, it would take twenty thousand years to finish.

One way to think of this data is as a poll of the entire book record. To get a sense of how comprehensive this poll is, consider that there are about as many registered voters in the United States (137 million) as the total number of books ever published (130 million). The Gallup poll released five days ahead of the 2012 presidential election surveyed 2,700 likely voters, about 1 in 50,000. Google's poll of all books includes 30 million books, or about 1 in 4. As polls go, it is incredibly comprehensive: an unprecedented précis of humanity's cultural record.

TWENTY-FIVE-YEAR-OLD GRADUATE STUDENT PSYCHOLOGY

Because we could not afford our own person-eon, it was clear that we needed to get in on the action at Google. But how?

Opportunity knocked when, in 2007, Aviva Aiden, Erez's wife, was invited to the Googleplex—Google's headquarters—to receive an award for women in computer science. Erez tagged along and made his way to the office of Peter Norvig, Google's famed director of research.

Norvig is a pioneer of artificial intelligence. He wrote the standard textbook on the subject. And when he talks, people listen. Many, many people listen. For instance, in the fall of 2011, Norvig and Sebastian Thrun taught the world's first massive open online course, or MOOC. Presented under the auspices of Stanford University, their artificial intelligence course was a runaway success: More than 160,000 students enrolled. It set off a revolution in higher education.

That makes his approach to meetings surprising. Norvig does not like to say much. In fact, the only thing harder to read than Google's digital books collection is Norvig's impenetrable poker face as he listens to you talk. Finally, after some time, he typically says something that is either very insightful or a complete non sequitur. With that, you know if you've succeeded in making your case.

After listening to Erez present our hour-long pitch, Norvig finally showed his cards.

"This all sounds great, but how can we do it without violating copyright?"

FORTUNE 500 LEGAL DEPARTMENT PSYCHOLOGY

When Google publicly announced its intention to digitize all the world's books in 2004, the publishing industry became—understandably—nervous. What would it mean for them if their books were to become searchable on the Web? Which content did Google intend to share with the public? Even if Google wanted to obey copyright law, how could the company figure out who held the rights to any given book? Would Google just overthrow the whole industry, as Apple's iTunes had done with music?

Soon, lawsuits began to pour in. On September 20, 2005, the Authors Guild, representing a huge number of individual authors, filed a class-action lawsuit. By October 19, the American Association of Publishers, representing megapublishers McGraw-Hill, Penguin USA, Simon & Schuster, Pearson Education, and John Wiley, filed its own lawsuit. Both suits alleged "massive copyright infringement." In 2006, French and German publishers joined the fray. By March 2007, Google's competitors were piling on, too. Thomas Rubin, one of the top attorneys at Microsoft, delivered a set of prepared remarks blasting Google's effort at digitization, saying that Google's approach "systematically violates copyright" and "undermines critical incentives to create." The Google Books project was rapidly becoming one of the most important legal flashpoints in the history of big data.

Google Books' troubles are a harbinger of the legal challenges that big data research will face going forward. The most interesting big datasets are frequently in the hands of massive corporations—the Googles, Facebooks, Amazons, and Twitters of the world. In the hands of, but not necessarily owned by. The data typically comes from individual people, whether it's because they wrote a book, put up a Web page, or took a picture. Those people retain significant rights over the data—as well they should, since it is their creation. These rights can take the form of copyrights, or privacy rights, or intellectual property rights, or a litany of other rights. So the data isn't public, but it isn't private, either. Instead, it comprises a shared digital commons, a no-man's-land in which millions of people may have an interest, no entity has complete authority, and legal status is often obscure.

For scientists, this is a game changer. We have gotten used to a world in which we generate or obtain data and then analyze it however we want. At most, a scientist might need to get approval from an ethics panel. But the traditional approach would make each of the big data studies we mentioned in our introduction—from Levin's analysis of eBay to Barabási's study of cell phone movements—illegal and unethical. In the world of big data, the notion of getting everything and analyzing it later is a practical and moral impossibility. How can we take advantage of big data, if no one is willing—or even has the right—to hand it over?

Norvig's question had zeroed in on the crucial issue.

BIG DATA CASTS BIG SHADOWS

Asking Google to just hand us the full text of the world's books was going to be a nonstarter. Fortunately, we didn't need to.

That's because big data casts big shadows. Just as a shadow is the dark projection of a real object—a visual transformation that preserves some aspects of the original object while filtering out others—shadow data preserves some, but not all, of the original information. Though shadowing is more art than science, it's crucial to making progress when working on big data. The wrong shadow can be ethically dubious, legally intractable, and scientifically useless. But if you choose exactly the right angle, it's possible to obscure the legally and ethically sensitive parts of the original dataset while retaining much of its extraordinary power.

If you're very lucky, shadowing a dataset can be easy. For instance, often the problem with a big dataset is that it exposes sensitive personal information. If so, erasing the name of the person associated with each record seems like it ought to be enough. But it's rarely so simple. Trouble is, many big datasets are so information-rich that attaching a name to each record is, on closer examination, redundant: The record itself contains so many identifying characteristics that there's only one person on the planet it could describe. In such a case, removing the name doesn't accomplish much.

America Online learned this the hard way in 2006, when, in what was meant to be a magnanimous contribution to scientific research, it publicly released the search logs of more than 650,000 users. Of course, AOL redacted the logs: People's names were not included in the release, and each user's handle was replaced with a nondescript numerical value. AOL thought this would protect users' privacy. But AOL was badly mistaken.

By examining the now-public search logs and cross-referencing them with other widely available data, it was possible for people like *New York Times* journalists Michael Barbaro and Tom Zeller, Jr., to deduce user identities. Days after the data was released, Barbaro and Zeller noticed that, amid hundreds of other queries spanning a three-month period, user 4417749 searched for "landscapers in Lilburn, GA" and for many people whose last name was "Arnold." A quick look through the phone book suggested that the user was probably a sixty-two-year-old lady living in Lilburn named Thelma Arnold. When Barbaro and Zeller contacted Ms. Arnold and read her some of the queries from her own search log, she was flabbergasted at what AOL had done: "We all have a right to privacy. Nobody should have found this all out."

AOL realized its mistake and tried to rectify the problem. Only three days after releasing the data, the company took it offline. It also apologized, fired the researcher who released the logs, and fired the researcher's supervisor. A few weeks later, AOL's CTO resigned. But it was too late: The data had already spread across the Web. Because of its high-minded but poorly executed effort to catalyze research, AOL was hit with a wave of well-deserved negative publicity and a class-action lawsuit. The debacle became a classic example of how hard it is to anonymize big data—and, to those in the industry, a cautionary tale of the dangers a company can face when it wades into altruistic data sharing. AOL stood to gain almost nothing by releasing those logs, and in the end it paid a great price. Norvig had this, too, in the back of his mind.

Of course, names are not the only thing that can make a dataset compromising. Google Books has the opposite problem. One of the few pieces of a book's text that you can usually release without fear of a lawsuit is the name of its author. The rest of the book's text is protected by copyright.

How can big shadows help us navigate this impasse? To make use of big data, one needs to find a shadow that satisfies four important criteria. First, the shadow needs to protect the rights of the millions of people whose collective efforts created the original dataset. Second, it needs to be interesting. Third, it cannot run counter to the purposes of the company, which serves as the data's gatekeeper. Fourth, it needs to be something one can actually generate in practice. AOL's problem was not that it had released data about user searches; the problem was that the shadow it chose obscured far too little, and led to an egregious violation of our first criterion. When Jeremy Ginsberg created Google Flu Trends, he too released information derived from user searches. But his shadow aggregated the data in such a way that no one was harmed—except for the influenza virus.

Using big shadows gives us a way to protect the information in a dataset while still putting it to work. And it's not just the researchers involved who stand to benefit. Because an ideal shadow is ethically and legally innocuous, it's often possible to persuade its wary keepers to release it into the public domain. Thus, big shadows give us a way to transform highly guarded datasets into formidable public resources, usable by anyone with a bright idea, whether it's a scientist, a humanist, an entrepreneur, or a high school student. When we're talking to companies, we like to present this as a form of data philanthropy: Donating bits can be just as good as donating bucks, and it is, by definition, cheaper.

IN THE SHADOW OF GOOGLE BOOKS

For simplicity, think of the raw data of Google Books as one long table containing the full text of each book, coupled with information about the work, such as the book's title, the author's name and date of birth, the library of origin, and the date of publication. What big shadows are cast by Google Books? Many. But not all are equally promising.

One shadow consists of only the title of each book. This shadow includes about one hundred million words. The data is tiny in comparison to the full collection, and too small to enable much new science. But it's still quite problematic: Google considers these titles to be business intelligence, because the company doesn't want competitors to know which books it has scanned and which it has not. So the titles don't make for a good shadow.

Another shadow is the full text of all public-domain books—all books on which the copyright has expired. This is a really interesting dataset, potentially free of the thorny issues involved when there are rights holders. But it has two drawbacks. First, since copyright extends for so long, few books published after 1920 are in the public domain. This means that the periods in which the big data is by far the biggest—the twentieth and early twenty-first centuries—are almost totally unrepresented. Second, the antiquated laws that govern copyright often leave the status of any particular book ambiguous. Such ambiguities affect a vast number of books in Google's collection. Because it's unclear which books should be included, this shadow can be surprisingly difficult to compute.

What to suggest to Norvig?

We thought back to Karen Reimer's *Legendary, Lexical, Loquacious Love*. Wouldn't the experience of leafing through Reimer's book, the way that word frequency revealed the hidden psyche of the story and of its author, be more interesting if the story was a big chunk of the historical record of Western civilization, and if the author was, more or less, everyone?

The more we thought about it, the more her alphabetical novel seemed to hint at a shadow that was both simple and beautiful,

beautiful, beautiful, beautiful, beautiful. Why didn't we just expose the word frequencies in Google books?

More precisely, our idea was to create a shadow dataset containing a single record for every word and phrase that appeared in English books. These words and phrases—the fancy computer science term is n-gram—include *3.14159* (a 1-gram), *banana split* (a 2-gram), and *the United States of America* (a 5-gram). For each word and phrase, the record would consist of a long list of numbers, showing how frequently that particular n-gram appeared in books, year after year, going back five centuries. Not only would this be extremely interesting, it seemed to us that it would probably be legally innocuous. Reimer was never sued for publishing an alphabetical version of someone else's novel.

But there was still one danger: What if a hacker figured out how to use the public data on word and phrase frequencies to reconstruct the full text of all the books? Assembling a massive text from tiny, overlapping snippets is not an obviously unreasonable strategy. In fact, an analogous method is the basis of modern genome sequencing—the approach used by scientists to read the DNA inside a cell.

To solve this problem, we relied on a statistical fact: You don't have to go far in any given book to bump into a unique formulation. For instance, the previous sentence was probably the only use ever of the phrase "bump into a unique formulation," or at least, it was, until this sentence came along. So we added a simple fix: Our shadow would not include frequency data for words and phrases that had been written only a handful of times. With this modification, reconstructing the full texts would be mathematically impossible.

The resulting shadow—the ngrams—seemed extremely promising. The copyright protection the texts enjoyed would not be compromised at all (criterion 1). We knew from our work on irregular verbs, and from Reimer's novel, how much insight could be gleaned just by tracking the frequency of a single word (criterion 2). It would be a powerful new way to search for concepts, and thus an appealing notion for a company built on search (criterion 3). And counting words is possibly the simplest problem in computer science (criterion 4).

Of course, if we limited ourselves to ngram data, the words would be stripped of nearly all context, so we wouldn't be able to tell if someone writing about Elia Kazan was arguing that he was a great director or that he betrayed his friends by naming names during the Red Scare. But that's not a bug, it's a feature: The context was exactly what had made the data legally sensitive. Freed of context, we could make a strong case that our shadow dataset, and the tools it powered, could be shared not only with us, the researchers, but with the entire world. Our shadow hit the spot: It's the most fun you can have without breaking the law.

Ngrams were our answer. Norvig thought about this idea for a minute and decided it might be worth a shot. He helped us assemble a team: Google engineers Jon Orwant and Matt Gray, and our intern, Yuan Shen.

We were in. Suddenly we had access to the biggest collection of words in history.

LEADERS OF THE FREE WORD

Language is assembled out of words. But what is a word?

It's a weighty issue. Consider politicians. Throughout his career, President George W. Bush occasionally got creative with language, doing things like adding the prefix *mis-* before the word *underestimated*. These Bushisms made him the frequent butt of jokes and a punching bag for late-night TV. The language used by politicians is so carefully monitored that even something as seemingly minor as nonstandard spelling can be a hot "potatoe." In his memoirs, former vice president Dan Quayle described the experience of publicly misspelling *potato* as "more than a gaffe. It was a defining moment of the worst imaginable kind." Yet Sarah Palin, faced with public ridicule after she used the word *refudiated* in a tweet, pointed out that, like other politicians, she was being held to a double standard. After all, she tweeted, "English is a living language. Shakespeare liked to coin new words too."

And she's right. Shakespeare's plays are chock-full of neologisms. In fact, like Bush, Shakespeare was a social conservative and a prefix liberal. He often created new words using the same strategy that led Bush to create *misunderestimate*. But unlike Bush, Shakespeare got away with it, leaving a vast lexical legacy as his coinages became widely adopted. For instance, he used the prefix *lack-* to create words like *lack-beard*, *lack-brain*, *lack-love*, and *lack-luster* (a word whose subsequent career has been anything but). Poets in general enjoy more lexical leeway than politicians do. Lewis Carroll's poem "Jabberwocky" is composed mostly of words that Carroll made up—and he'd probably *chortle* at how many of them are considered proper English today.

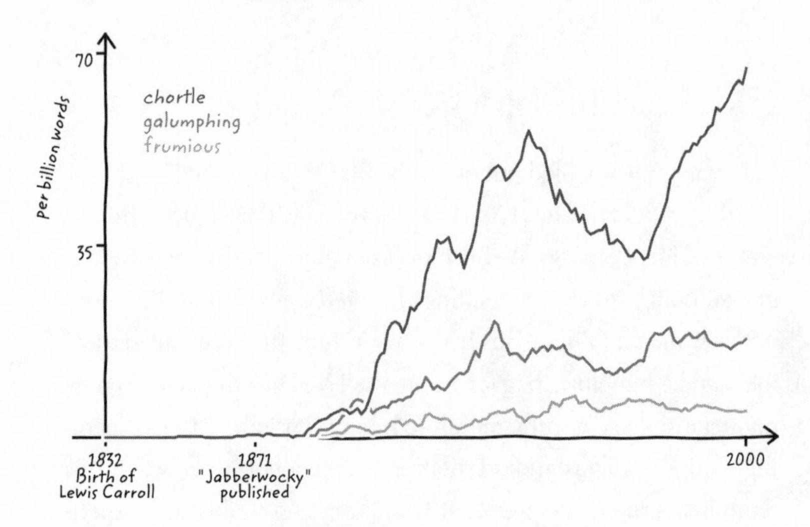

So how can we decide what words are okay to use, and which will transform us into a late-night punch line?

TO WORD, OR NOT TO WORD?

Lexicographer. A writer of dictionaries; a harmless drudge . . .

—Samuel Johnson, *A Dictionary of the English Language*, 1755

Dictionaries, at least in principle, solve the problem of what is or is not a word. After all, dictionaries are catalogs of officially approved words, each paired with a list of approved meanings. Many dictionaries are meant to be handy references, like the *American Heritage Dictionary*, whose fourth edition lists about 116,000 words. Some dictionaries are more ambitious, none more so than the comprehensive, twenty-three-volume compendium known as the *Oxford English Dictionary*. First completed in 1928, the most recent edition of the *OED* lists 446,000 words. If you want to know what words are officially part of the language, dictionaries are the place to go. If it's in the dictionary, it's a word. If it isn't, it's not.

But even if that's the case, we still have a puzzle on our hands. How exactly do the lexicographers who write dictionaries know which words to include?

There are two ideas about how this works.

One theory is that the lexicographer's job is prescriptive. According to this view, lexicographers are in charge of what is in the language, and in writing dictionaries, they legislate what words we should and should not use. This was Teddy Roosevelt's "Bull Moose" view of lexicography. In 1906, he ordered the Government Printing Office to begin using a drastically simplified spelling: *I have answered your grotesque telephone* became *I hav anserd yur grotesk telefone*. This did not go down well with Congress, and the original spellings remained untouched. The prescriptive view of lexicography is still dominant today in France, where every now and then the government publishes an official document about correct word usage and spelling. In January 2013, the *Journal Officiel* recommended that *hashtag* be replaced with *mot-dièse* (roughly, word–pound-sign). Of course, the Twitterverse responded with a collective #ROFL. The problem with the prescriptive approach is that it's not obvious that anyone is, or should be, in charge of language. A language transcends any particular government, ethnicity, or nationality.

A different idea—one that is more widely believed today, especially in the United States—is that the lexicographer's job is not *prescriptive*—telling us what to do—but instead *descriptive*—reporting what we do when left to our own devices. According to this approach, lexicographers are not monarchs but explorers. A dictionary is a map of what they have found.

But there's a problem with this idea, too. If lexicographers can't decide what a word is by fiat, then isn't it possible for them to make a mistake? How much can we really rely on the dictionary?

After all, lexicographers are ordinary people. Sure, they may be more interested in nuances of usage than the average person on the street. But when trying to figure out what words to include in their dictionaries, lexicographers typically do the same kinds of things the rest of us do. They listen to what people are saying. They read a lot. They try their best to notice trends: What new words are people using? What words have they stopped using? What entries are popping up in competing dictionaries?

Once they form those personal impressions and identify a candidate word, lexicographers try to figure out if those impressions are real. One lexicographer we know, when trying to decide if something is a real word, uses the following criterion: He tries to find four examples of that word in unrelated pieces of writing. Consensus among the lexicographic team is desirable, but technical jargon—the decision whether to include a word like *graphene*—may be left up to the judgment of the one consultant who handles physics. Writing dictionaries is not a science. It's a centuries-old art.

Take the *American Heritage Dictionary*. Its fourth edition was published in 2000, eight years after its third edition. In those eight years, new words had entered the language. The editors at the *AHD* did their best to hunt those words down. Their trophies included *amplidyne* (a type of power generator), *mesclun* (a type of salad), *netiquette* (Internet etiquette), and *phytonutrient* (the chemicals that give plants their color, flavor, and smell). How well did they do?

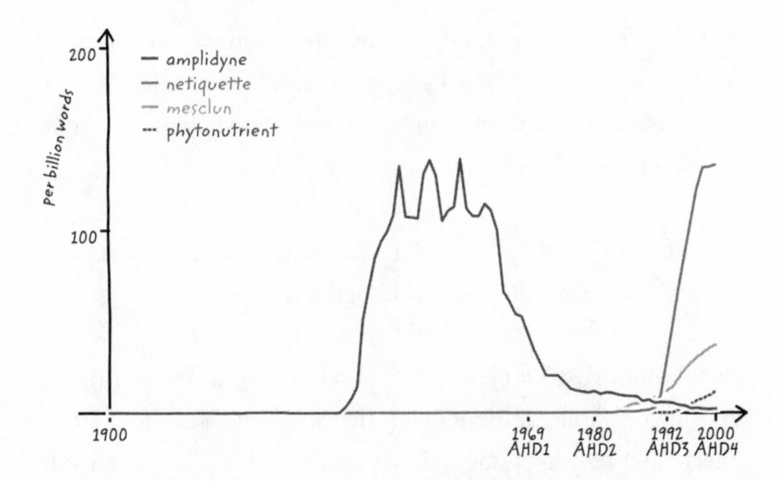

This graph makes it clear that the *AHD*'s record is mixed at best. In some cases, like *mesclun* and *netiquette,* they were merely late to the party. Purely on the basis of frequency, both words should have qualified for the *AHD* in 1992. In the case of *amplidyne,* the party was long over; amplidynes peaked in the early twentieth century and are completely obsolete today. Despite their best efforts, lexicographers are hard pressed to detect new words in time, and can be decades behind.

When we saw this plot, we knew that—at least when it came to identifying words—being able to read billions of sentences in one click could be a godsend to lexicographers.

DIY DICTIONARY

We decided to create our own descriptive lexicon containing all of the words used in contemporary English. Our idea was simple: If a string of characters is frequent enough in contemporary texts written in English, then it's a word. How frequent is frequent enough? The natural cutoff is to use the frequency of the rarest words in dictionaries, which we calculated was roughly one instance in every billion words of text. So our answer to the question "What is a word?" is:

> An English word is a 1-gram that appears, on average, at least once in every billion 1-grams of English text.

This obviously isn't a perfect definition of a word. Does "English text" include a Spanish quotation that might be embedded in an otherwise English passage? Does the text have to be recent? Should it come from books? Transcribed speech? The Internet? Should we really count common misspellings, like *excesss*, as words? What about partly numerical forms, like *l8r*? And can't a 2-gram, like *straw man*, actually be a word?

Yet for all its faults, this definition is actually pretty precise. It's precise enough that, furnished only with this definition, an agreed-upon reference text of sufficient length, and a bunch of computers, one can create an objective lexicon of the English language. In this one respect, our definition is better than the highly subjective formulations found in most references.

We wanted to make sure that our new, Zipfian lexicon represented contemporary usage, so we didn't just throw all our books at it. Instead, we took a ten-year slice of our data—all the books in our database that were published between 1990 and 2000. This collection contained more than fifty billion 1-grams. For a 1-gram to meet our cut-off frequency of one in a billion, it had to appear at least fifty times in the collection. The resulting list contained 1,489,337 words, like *unhealthiness*, *6.24*, *psychopathy*, and *Augustean*.

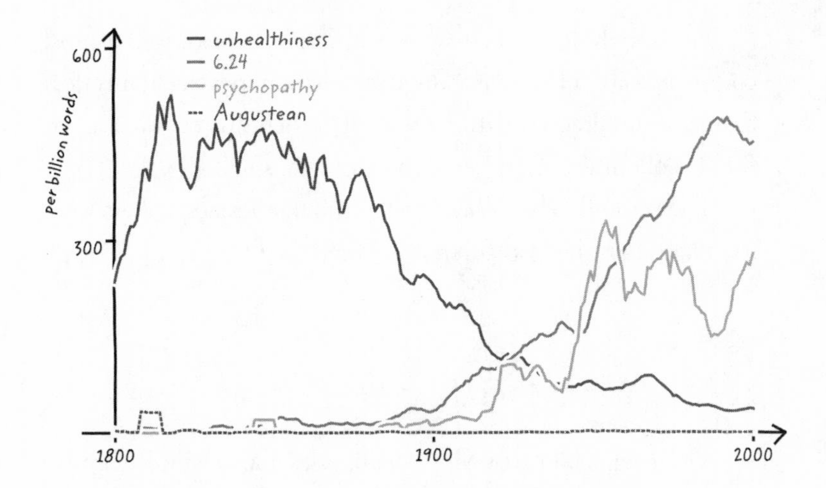

Our Zipfian lexicon is a pretty handy reference. If a word doesn't appear in this list, then it's not as frequent as the least frequent words in the dictionary—and it's pretty reasonable to argue that it isn't a word. If it appears, the word is probably frequent enough to warrant inclusion in the dictionary. If it is not included in dictionaries, one has to wonder why.

This is one of the fun things about having an objective lexicon. All these years, whether at school or at Scrabble, the dictionary has been used to test you. With an independent way of assessing the lexicon, the shoe is on the other foot, and it's possible for you to test the accuracy of the dictionary and the lexicographers who wrote it. There have been armchair lexicographers for centuries, but only with ngrams can one become an armchair *lexicographerologist*. (*Lexicographerology*: the study of harmless drudges. *Lexicographerologist*: an even more harmless drudge.)

Next, we asked the most fundamental question in all of lexicographerology: How much of our Zipfian lexicon did the dictionary catch?

Surprisingly little. The *Oxford English Dictionary*, the most comprehensive of English-language dictionaries, contains less than five hundred thousand words. Its lexicon is roughly a third the size of our list. All other dictionaries are smaller.

How could this be? Was it really true that lexicographers were so unaware of what was going on in their own language?

LEXICAL DARK MATTER

We have been a bit hasty. Most dictionaries don't claim to capture all words in the language. In fact, many dictionaries, on principle, are careful to exclude several types of terms, regardless of how common they might be:

1. Words that aren't entirely composed of letters (*3.14* and *l8r*)
2. Compound words (*whalewatching*)
3. Nonstandard spellings (*untill*)
4. Words that are hard to define (*AAAAAAARGH*)

As such, it's unfair of us to go "gotcha!" when we include things in our list that the dictionary isn't even trying to include. To get a sense of what the dictionary leaves out that it didn't deliberately mean to leave out, we estimated what fraction of our word list came from the above four categories.

That cut our list down from just under 1.5 million to a little over a million words. Still, our Zipfian lexicon had more than twice as many entries as the *Oxford English Dictionary*. Even the most comprehensive dictionary of the English language misses most words. These undocumented words included many colorful concepts, like *aridification* (the process by which a geographic region becomes dry), *slenthem* (a musical instrument), and, appropriately enough, the word *deletable.*

So, what trips up dictionaries?

Frequency. It turns out that dictionaries have excellent coverage of frequent words. Dictionaries are completely perfect—they literally contain 100 percent of all words—as long as those words are more frequent than one in a million, such as the word *dynamite.* If a word appears at least once in the average pile of ten books, the dictionary will record it and define it, like clockwork.

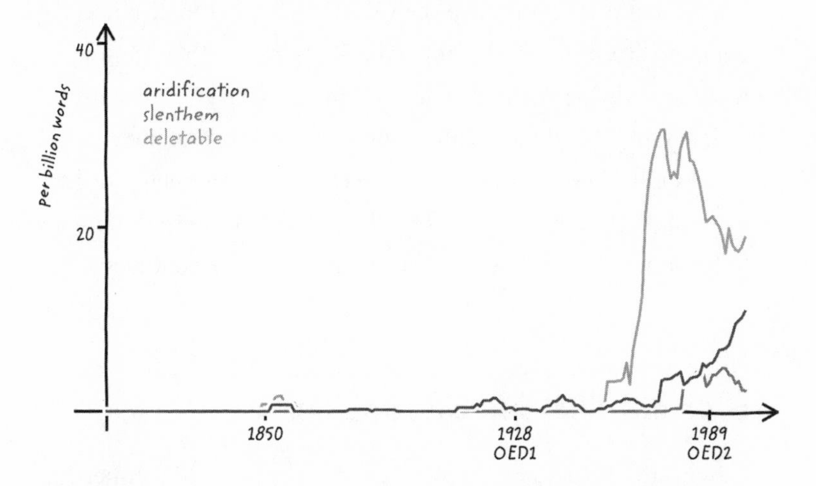

But lexicographers struggle with the rare stuff. As a word's frequency drops below one in a million, the chances that a dictionary has omitted the word will skyrocket. At frequencies just north of one in a billion, the dictionary only notices a quarter of all words.

And if there is one thing you should remember from Zipf, it is that most words are really rare. So, if dictionaries miss most rare words, then they miss most words, full stop.

As a result, it turns out that 52 percent of the English language—most of the words used in books—is lexical dark matter. Like the dark matter that makes up the majority of the universe, lexical dark matter makes up the majority of our language, but goes undetected by standard references.

As the limitations of traditional lexicography have become increasingly apparent, the field has begun to change. New entrants, like wordnik.com, wiktionary.com, and urbandictionary.com, have come to rely on armchair lexicographers in their efforts to build comprehensive online dictionaries. In effect, they are attempting to use crowdsourcing to document all of the dark matter. Traditional dictionaries like the *OED* are hoping to dive into big data, too. To bring their compendia up to speed, they are supplementing existing methods with an emerging style of data-driven lexicography. (And even with a touch of lexicographerology!)

Overall, these developments are certainly good news for lexicographers. Despite centuries of effort, most of the work remains to be done. English is, by and large, an uncharted continent.

FOUR BIRTHDAYS AND A FUNERAL

New words always get people excited. Every year, the American Dialect Society holds a meeting to honor all these new words. Members vote on categories like Word of the Year, Most Outrageous, and even Least Likely to Succeed, a distinction that our own coinage—culturomics—earned in 2010. Since 1991, Words of the Year have included *cyber* (1994), *e-* (1998), *metrosexual* (2003), and most recently *hashtag* (*mot-dièse,* in case the French government is reading). The lists compiled by the American Dialect Society testify to a language that is constantly welcoming and celebrating new words.

But there's very little activity at the other end of the lexical life cycle. Nobody seems interested in holding funerals for words that have died. So it's hard to tell whether the birth rate exceeds the death rate—whether English is growing, shrinking, or remaining stable.

To find out, we created two more Zipfian lexicons. The first time around, we had used texts published between 1990 and 2000 to create a contemporary lexicon. This time, we used two historical periods: the decade preceding 1900, and the decade preceding 1950.

We found that by 1900, the lexicon already contained more than 550,000 entries. That's more words than are in today's *Oxford English Dictionary.* For the next fifty years, not much seemed to happen, and the language remained stable in size. Births and funerals managed to hold serve.

But between 1950 and 2000, English entered a period of growth, nearly doubling in size as hundreds of thousands of new words were added. New births dramatically outnumbered lexical last rites. Currently, about 8,400 words enter the English language each year—more than 20 new words crossed the threshold today.

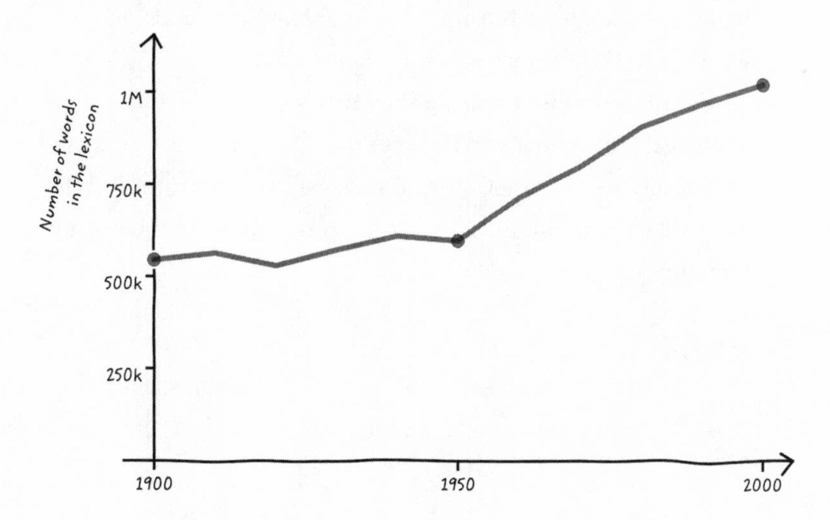

Our language is not only changing—it's growing.

Why is that? Nobody really knows, although, as with the cause of power laws, conjectures abound. One hypothesis is that as our society becomes increasingly connected (we keep in touch with more people) and our world gets smaller (people are at most a phone call or a plane ride away), new words reach critical mass more easily. Another hypothesis suggests that progress in science, medicine, and technology introduces new words as jargon enters the general parlance. Yet another possibility lies in diversification within the book record itself, the basis of our Zipfian lexicon. As a broader cross section of society began to publish books in the late twentieth century, authors wrote about more topics in a wider range of dialects, introducing more words to the global discussion.

Truth be told, no one knows for sure. And since we don't know where this effect is coming from, it is hard to guess where it is all going. Will the number of words born each year increase? What is the limit on the size of the lexicon? How different will your kid's language be from yours? As the scopes of big data illuminate our language, they light the way to a new scientific landscape, one where even the Sasquatch has nowhere to hide.

But the words we use tell a story much greater than that of our language. They are a window into our thoughts, our mores, and our society itself. So let's turn our scope away from the mechanism of our communication, and toward the substance of our thought.

Daddy, where do babysitters come from?

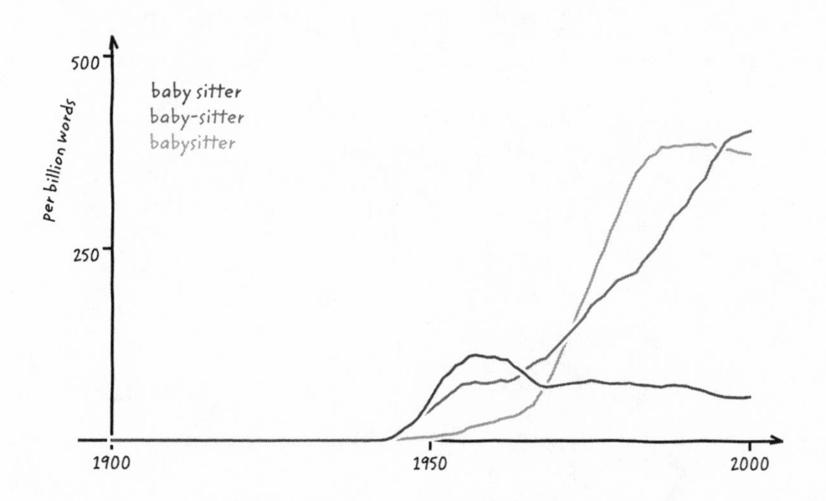

In the mid–twentieth century, it turned out that taking care of a *baby* using a *sitter* was a very good idea. Since the words *baby* and *sitter* had such compatible interests, they started spending a lot of time together, and *baby sitter* became increasingly frequent.

Soon, people started to view them as joined at the hip. They represented this join with a hyphen. As the relationship got more serious, *baby-sitter* became increasingly frequent, and *baby sitter* started to get replaced.

Eventually, *baby* and *sitter* realized they were a match made in heaven. A child was born of this union. And that, dear child, is why your parents left you here with me, the *babysitter*.

4

7.5 MINUTES OF FAME

Getting rid of crap is not sexy. But it can be heroic.

Just ask Hercules, the hero-god of Greek mythology. For his fifth of twelve labors, Hercules was tasked with cleaning the Augean stables, which housed thousands of immortal cows. Because the stables had not been cleaned in thirty years, they had come to contain a sizable cache of waste. Hercules redirected two raging rivers to purge the stables in a single day. His heroic deed remains one of the greatest achievements in the annals of scatological engineering.

Millennia from now, similar legends will surely be told about Yuan Shen, our own computational Hercules. Google had spent five years grazing at the rich pastures of world knowledge, its swift scanning process ingesting books by the million. Yet as an inevitable by-product of having created the world's largest stable of digitally immortalized books, the company had accrued a significant quantity of poop-grade data as well. Big data is messy. The time had come to clean the stable.

CUT THE CRAP

How much quality time have you spent with a library card catalog lately?

Card catalogs used to be the heart of library circulation. There was one card for every book in the library, containing vital facts like the title, the author, the subject, the year of publication, and the all-important call number, which indicated where the book was located. Library visitors would stream into the card catalog all day long, and the information in the catalog would, in turn, pump them into the farthest corners of the stacks.

Without its card catalog, a library becomes a cluttered desk the size of a building: You can't find anything. For many centuries, one of the most important libraries, the Archivio Segreto Vaticano (the Vatican Secret Archive), was just this way. It lacked a comprehensive card catalog for the works that occupy its fifty-two miles of shelf space. What was in there? Even those who had unfettered access could answer only with a mixture of fact, rumor, and legend. Finding a book was a matter of knowing someone who knew someone who knew (or thought they knew) where the book was. The archive contains priceless manuscripts dating all the way back to the eighth century—like records from Galileo's heresy trial—but finding these treasures could be an adventure worthy of Indiana Jones. That's certainly one way of keeping a secret.

For us, like any other library users, access to the books alone was not nearly enough. If we wanted to compare texts from different times and places, we needed accurate card catalog metadata telling us what each book was, so we would know how to classify it in the context of an automated analysis.

Going in, we didn't think that was going to be a big problem: Google had assembled its shopping list of 130 million books using catalog information from hundreds of sources. (These days, the card catalogs of the major libraries have been computerized—one of the first things to benefit from digitization—and the physical cards themselves are often relegated to a side room.) But it turns out that card catalogs, even the best ones, are riddled with errors.

Once made, these errors don't get corrected very fast. There are so many cards, and even the most enthusiastic library users don't always notice the mistake. Either the error prevents a user from finding the card (in which case, "See no evil, hear no evil, speak no evil"), or the error lies in something like the place of publication. As long as the call number is still accurate, the user finds the book anyway. The problematic metadata on the card doesn't bother the reader much, because the correct information is already waiting on the book's title page.

Over time, those legions of uncorrected errors made their way from physical card catalogs to digital card catalogs, then to Google's mother of all catalogs, and then to us. Unlike people who are interested in reading a single book, we were particularly vulnerable to errors: We couldn't afford to manually look through each of the millions of books. Yet a large fraction of the cards contained mistakes. When we used this catalog metadata to produce ngram tables, the results were often so badly scrambled as to be unusable. According to our initial calculations, our friend in the office next door had enjoyed a surge in popularity during the sixteenth century. When we confronted her about this, she denied being that old. Either she was lying to us, or we had a very big problem on our hands.

What to do?

Since we couldn't go through the books by hand, we decided to write computer algorithms to look for suspicious-looking cards—for anything that suggested that the information on a card might be erroneous. For instance: magazines. Libraries typically assign every single issue of a serial publication—be it a newspaper, an academic journal, or any other periodical—the publication date of the very first issue. That means every issue of *Time* magazine was, according to our card catalog, published in 1923. For our purposes, this was a very big problem.

To resolve these issues, we wrote an algorithm called the Serial Killer to find anything that looked like it might be a serial publication. Another algorithm, called the Speed Dater, looked at a book and tried to guess when it was published based on the text it contained. Together, these approaches helped us identify suspicious cards and the books that they belonged to. We could then exclude these books from our analyses.

MR. CLEAN

Finally, in the summer of 2009, Yuan combined these methods with his software engineering muscles in order to wash away the crap that was befouling our big data. Millions of books were flushed in a river of computation so massive that it set off Google's internal warning systems. What was left after this laundering of legendary proportion was only a fraction of what we had started with. Nevertheless, it was still unprecedented in size and historical depth: five hundred billion words, written over five centuries, in seven different languages. It contained more than 4 percent of all books ever published.

Just as important, the massive dataset gleamed. Despite the fact that the total amount of text was a thousand times longer than the human genome, it was—letter for letter—ten times as accurate as the sequence reported by the Human Genome Project.

And now that the input texts and the card catalog metadata were pristine, the ngram data they produced looked great. We could clearly discern a vast array of linguistic and cultural changes, like the shift from *throve* to *thrived*, and the progression from *telegraph* to *telephone* to *television*. As soon as we caught a glimpse of the ngram data, it was, scientifically speaking, love at first sight.

But like so many summer romances, our love affair with ngrams would face obstacles come fall. With Yuan's internship wrapping up at the start of the academic year, we would soon find ourselves back outside Google, leaving our data behind the company's firewall.

We needed Google to send us the data. But the Internet giant didn't want to. By Google's account, ngram data was still extraordinarily sensitive. The ngram dataset had been calculated from the full text of five million books, and Google's legal calculus was simple. Five million books corresponds to five million authors, which corresponds to five million plaintiffs in the massive lawsuit that might result if the data were to leak. We had specifically designed the ngram shadow dataset to get around this problem by counting words instead of recording long stretches of text. But our combinatorial sleight of hand had not yet been tested in a court of law. Google was understandably wary.

We had very few cards to play when faced with the legal department of one of the world's largest corporations. But with two billion ngrams in the pot, we were not yet ready to fold.

WHAT FAME BUYS YOU

We had exhausted one card after another. Chance, in the form of Aviva Aiden receiving an award, which initially opened the doors of the Googleplex to us. The kindness of strangers, in the form of Peter Norvig's green light and his willingness to collaborate. We had even "phoned a friend," when a long-lost neighbor, Ben Bayer, turned out to be the "Master of Space and Time" at Google Research (possibly the greatest job title in corporate history). But there was one card we had yet to play.

All our talk about quantifying historical trends had caught the attention of Steven Pinker, one of the most prominent scientists alive today and someone whom we had always admired.

Pinker is a psychologist, linguist, and cognitive scientist of extraordinary breadth and depth. The author of numerous bestsellers, he has the uncanny ability to distill the most complex problems to their very essence in a crystal-clear way. For instance, on one occasion, Pinker appeared on the satirical news show *The Colbert Report*. Stephen Colbert asked him, "How does the brain work? Five words or less." Pinker thought for a couple of seconds and said, "Brain cells fire in patterns."

As luck would have it, one of Pinker's fans is none other than Dan Clancy, who in the summer of 2009 was the head of the entire Google Books operation. Clancy was high enough on the totem pole that his word alone would be enough to get us off-campus access to the ngram data. But Clancy is a busy, important guy who had no time for the likes of us or our little project. Still, as the summer drew to a close, it became clear that if Pinker would be willing to show up for a meeting to discuss the ngrams, then the elusive Dan Clancy would find the time to make it, too.

So we asked Pinker: Look, we've generated these two billion ngrams—could you help us liberate them? Pinker thought our work had the potential to be useful and agreed to come. So Clancy agreed to come, too. We had thirty minutes to make our case.

Some years ago, Pinker had been named one of the hundred most influential people on the planet by *Time* magazine. As the meeting got under way, it was clear why. Half an hour was more than enough time for him to work his magic. Soon, the ngrams were on their way.

So what does fame buy you? Pinker's fame bought us thirty minutes of Clancy's time. Not much—but it was enough.

THE STORY OF FAME

> Fame is a bee.
> It has a song—
> It has a sting—
> Ah, too, it has a wing.

This poem by Emily Dickinson captures the essence of fame: the allure, the danger, the way it elevates a person, and its tendency to float just beyond our reach. Dickinson, one imagines, should know. She is perhaps America's most famous poet.

Yet Dickinson's relationship with fame is not straightforward. What she knew about fame she knew from intuition, not experience. A complete unknown during her lifetime, Dickinson left behind poetry that became the subject of widespread discussion nearly half a century after she died in 1886.

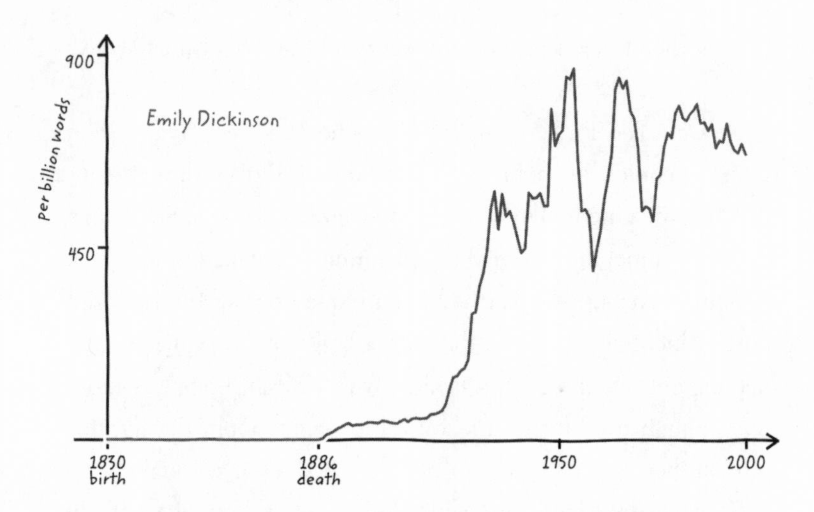

Is Dickinson's relationship with fame the exception or the rule? Fame finds people in so many different ways, at so many different times, and for so many different reasons that there seems to be no typical route. Prince William, son of Prince Charles and Princess Diana, was famous from the very moment of his birth, or even earlier, given that his destiny to become the king of England was preordained from the womb. Pop singer Justin Bieber was discovered on YouTube when he was only thirteen; five years later, Bieber was the most Googled person on Earth. Sometimes, a lifetime of learning translates into overnight fame, as when Pinker, already an MIT professor, soared to worldwide acclaim at age forty with the publication of his runaway bestseller *The Language Instinct.* On the other hand, Julia Child didn't start learning to cook until she was past forty. But that still left her with enough time to revolutionize American cuisine and become a national icon.

Like Emily Dickinson, many of the most famous people never experience fame in their own lifetimes. Almost none of Vincent van Gogh's paintings sold during his lifetime; he died with his genius unrecognized. The monk Copernicus understood that his big idea—the notion that the Earth circled the sun and not the other way around—was so incendiary that he waited until he was on his deathbed to see it published. In some lines of work, posthumous fame is the norm. As Union general William Tecumseh Sherman put it, "I think I understand what military fame is; to be killed on the field of battle and have your name misspelled in the newspapers."

And then there are the people who appear to be famous for no particular reason at all. Famously famous folks, like Paris Hilton and Kim Kardashian, develop a reputation for being famous that can become a sort of self-fulfilling prophecy. Such people highlight the extraordinary gravitational pull that fame exerts: It is not only the achievements of famous people that draw us to them, but the very fact that they are famous, in and of itself.

Given how fascinated we all are with fame, it's quite surprising how little we understand about how it works.

THE WRIGHT STUFF

What is fame? Like energy or life, fame is an everyday concept that we all intuitively grasp but find extremely hard to define. (When Justice Potter Stewart famously said of pornography, "I know it when I see it," he could just as well have been talking about fame.) It's also clear that fame comes in a wide variety of sizes: Everybody knows that Jesus is more famous than singer John Lennon, that Lennon is more famous than actor Alec Baldwin, and that Baldwin is more famous than hot dog–eating champion Takeru Kobayashi. But again, a precise definition of what it means to be "more famous" is hard to come by. Like love and beauty, fame is hard to define, and harder still to measure. Yet if we hope to understand fame, learning how to measure it would be invaluable. Measurement, although not the solution for all intellectual problems, is a great tool for demystifying notions that might otherwise remain ambiguous and flighty.

Take the concept of flight itself. In 1903, thanks to the recent development of automobiles, aeronautical engineering was all the rage. There were no garages back then (the ngram for *garage* is virtually nonexistent prior to 1906), but if there had been, every one of them would have been filled with an inventor scrambling to build the first airplane, a heavier-than-air device that could take off under its own power and engage in controlled flight. Existing machines didn't fit the bill. Either they couldn't get off the ground or they crashed immediately. Most inventors believed that the problem was the engine. If only they could make an engine powerful enough, they could achieve the dream of flight.

But Orville and Wilbur, two bicycle mechanics from the Midwest, didn't see it that way. The Wright brothers thought that the real problem was waiting in the wings. If you didn't have a decent wing, they reasoned, a better engine wouldn't help. At the time, there were already extensive mathematical theories about how wings should perform. But when the Wrights studied the theory, they realized that it didn't match up with what they were seeing in their failed test flights. When it came to wings, they decided, theorizing could only take you so far. The theory made underlying assumptions about the physical world, and those assumptions might be wrong. So the problem was not one of theory, but of measurement. What they needed was a way to study the aerodynamics of airplane wings experimentally—to create test wings and to rapidly measure how well they worked.

So, amid intense competition, the Wright brothers took a calculated risk. Instead of plowing ahead with more flight tests, they holed up in the back of their bike shop in Dayton, Ohio. There they spent months building a precise measurement tool for wing performance. The result was a small gasoline motor creating constant airflow through an adjacent six-foot-long wooden chamber: a wind tunnel. Using their wind tunnel, the Wrights could quickly measure one wing design after another, precisely ascertaining how much lift and drag each airfoil produced. Of course, their measurements of the performance of airfoils in a wind tunnel were a simplification, an imperfect simulacrum of the actual performance of an actual wing on an actual plane in actual flight. But, they reasoned, data is better than no data. If your aeroplanes keep crashing, it's better to introduce some sort of measurement than to rely on intuition, moxie, and a good fire extinguisher.

Their bold move turned out to be crucial, enabling them both to patch up the theory and to go beyond it. As Wilbur Wright later recalled:

> It is difficult to underestimate the value of that very laborious work we did over that homemade wind tunnel. From all the data that Orville and I accumulated into tables, an accurate and reliable wing could finally be built. As famous as we became for our "Flyer" and its system of control, it all would never have happened if we had not developed our own wind tunnel and derived our own correct aerodynamic data.

It turned out that the Wrights' wind tunnel—albeit simple—was not too simple to capture the important aspects of what made for a good wing design. In their tunnel, the brothers could precisely measure the performance of one airfoil after another. Based on the resulting data, they built a highly optimized wing and slapped it onto a plane. On the morning of December 17, 1903, they entered history, flying.

If we want to understand fame, what we need is a wind tunnel.

ALMOST FAMOUS

Many aspects of fame are difficult to measure. The loss of anonymity. The pressure of the spotlight. The psychological impact of watching your star wane.

But what about the bigness of fame—that sense that Jesus is more famous than Lennon, who is more famous than Baldwin, who is more famous than Kobayashi? Here, perhaps, there is hope. After all, an important aspect of the magnitude of fame is how frequently people mention you. And an important aspect of how frequently people mention you is how frequently people mention you in books. And when it comes to mentions of people in books—well, ngrams can really come in handy for that.

Of course, what we measure with ngrams is not fame itself but a simplification, a fame facsimile. Let's call it "phame" for now. The question is, does phame resemble fame well enough to serve as our wind tunnel?

Let's start exploring this question with a look at Charles Dickens, one of England's most famous writers. His first novel, *The Pickwick Papers*, began in 1836 as a serial—a book published in a periodical as a series of small parts. With the publication of *The Pickwick Papers*, the 2-gram *Charles Dickens* begins to pick up speed in the book record. Like the Wright brothers' famous *Flyer*,

Dickens' phame just kept on rising as he produced a steady stream of bestsellers, including *Oliver Twist* (1837), *A Christmas Carol* (1843), *David Copperfield* (1849), *A Tale of Two Cities* (1859), and *Great Expectations* (1860). The cultural impact of these works was enormous. It is said that *A Christmas Carol* popularized the greeting *Merry Christmas*, a report that is consistent with the ngram data.

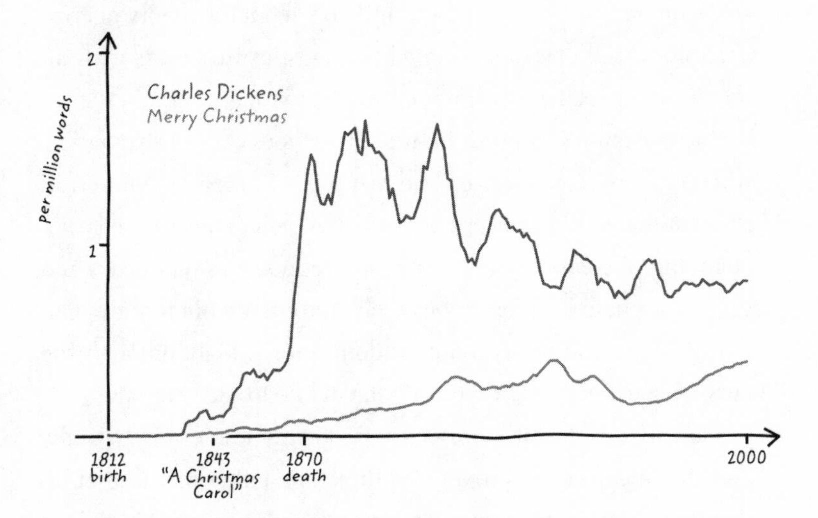

As with Dickinson, Dickens' death in 1870 did not cause his phame to ebb. Instead, it skyrocketed, as word of his passing brought on a newfound appreciation for his genius. In the decades after his death, his frequency of mention reached its very peak. But by 1900, the 2-gram *Charles Dickens* had begun a slow decline. Despite being extraordinarily phamous even today, the subject of intense scholarly examination, and a staple of high school curricula, Dickens' phame is plainly on the wane. It has been for over a century.

Putting *Charles Dickens* into our wind tunnel produced interesting results—a plausible measurement of the public interest that resulted from Dickens' achievements.

But the outlook is not completely rosy. Our example also helps highlight some of the important ways in which phame, as measured using books, and fame, as reflected in our intuitive notions of cultural importance, don't always get along famously. All measurement devices make mistakes. To better understand what's going on here, it helps to know a little bit about the theory of error analysis, a well-developed branch of statistics that deals with all the ways in which a measurement can go wrong.

Statisticians distinguish between two types of error that a measurement device can make. The first type is called random error: fluctuations that occur even if what is being measured is not changing. We can see such errors in the form of small peaks and valleys in phame, which, though ubiquitous, are often not meaningful. The good thing about random error is that, although the curve wiggles around, it typically stays close to the true value.

So-called systematic errors are trickier. These errors typically skew the measurement in a given direction, either inflating or reducing it. For instance, our procedure for measuring phame is to search for instances of a person's name. But this captures only a small fraction of all references. If we're tracking the frequency of *Charles Dickens*, we miss cases in which people refer to him as just "Dickens" or "Charlie" or "C-Money." If they refer to him as "the author of *The Pickwick Papers*" or "the husband of Catherine Hogarth," we won't catch it, either. And of course, if someone makes a reference to Dickens' legacy by quoting a favorite passage, or admiring a trick by illusionist David Copperfield, or even just using the phrase *Merry Christmas*, we miss that, too.

A great example of the difficulty involved in catching every single Dickens reference occurred when Michael Steele, running for chair of the Republican National Committee, was asked to name his favorite book during a televised debate in 2011. Steele's answer was an embarrassing gaffe: "*War and Peace* . . . the best of times and the worst of times." The quote is mangled Dickens, from *A Tale of Two Cities*. But *War and Peace* is by Leo Tolstoy. Was Steele referring to Dickens or wasn't he?

These types of errors—when we neglect something we'd ideally want to catch—are a class of systematic error that statisticians call a false negative. As a result of our false negatives, the phame we report is typically much lower than the true frequency of references to a person.

There is another type of systematic error, called a false positive. This occurs when we count something that we really should not. Someone writing the words *Charles Dickens* may in fact be referring to Dickens' eldest son, the author Charles Dickens, Jr.; his grandson Gerald Charles Dickens; two of his great-grandsons, Cedric Charles Dickens and Peter Gerald Charles Dickens; or his great-great-grandson, the actor Gerald Charles Dickens. Phame chalks it all up to the family patriarch. But statisticians know that this can be perilous. No statistician understands this issue more deeply than a professor at UC Berkeley named Michael I. Jordan. To see why, Google *Michael Jordan statistics*.

But we've yet to broach the most complex statistical issue raised by our technique.

Consider the year 1936. Many famous people were born in 1936. Two of them are Robert Redford and Václav Havel.

Robert Redford is the quintessential Hollywood star. He has played iconic roles in films for the last five decades, inspiring hundreds of millions of people with his performances in movies like

Out of Africa, *The Sting*, and *All the President's Men*. His rugged good looks have made him one of America's best-loved cultural figures, known the world over.

Václav Havel is a different breed of celebrity. He was a quiet playwright who led Czechoslovakia out of communism during the Velvet Revolution, becoming its first president. Four years later, he presided over the peaceful separation of the Czech and Slovak republics. Havel is one of the most famous political and literary figures of the twentieth century.

Both of them are among the ten most phamous people born in 1936. But they are edged out for the spot at the top of the list. Who, then, is the most phamous person born in 1936? A woman named Carol Gilligan.

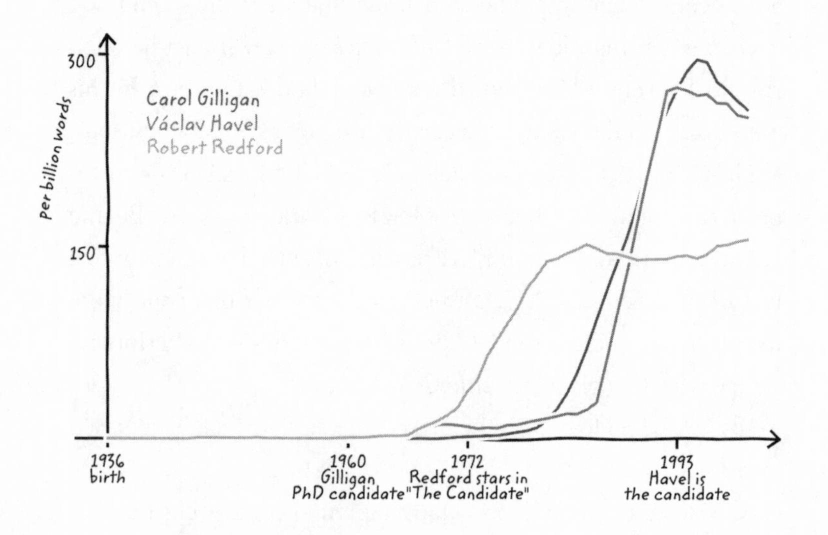

Gilligan is a renowned psychologist and a prominent feminist, whose groundbreaking work has led to positions at Harvard, Cambridge, and now New York University. Like Pinker, she's been on

Time's list of the most influential Americans. She is an intellectual superstar. Books mention *Carol Gilligan* a whole lot, a bit more often than either *Václav Havel* or *Robert Redford*. If phame and fame were exactly the same, then the most famous of all would be the scholarly dame.

But let's get real. Carol Gilligan is not more famous than Robert Redford. She's talked about more in books, because she's exactly the type of person that the type of person who writes books tends to think about: a science celebrity and a social critic. But she's not the type of person who makes headline news every day, not the type of person whose image is likely to pass by on the side of a bus, and not the type of person who makes teenage girls fawn by the millions.

The problem is that phame doesn't capture this bigger picture. If you were to take into account mentions on TV news, mentions in tabloids, mentions on Internet celebrity sites, and mentions around the office water cooler, Havel and Redford surely eclipse Gilligan, and by no small margin. Gilligan is benefiting from what statisticians call sampling bias—the aspect of culture that phame measures gives her an unfair advantage. She is more phamous than she is famous.

Our wind tunnel is not without its flaws. But these faults are not unique. Instead, they fall into classic error categories that arise with any measurement tool and that scientists and statisticians have been dealing with for decades. Bearing these imperfections in mind will make it possible to develop better tools in the future.

The relationship between phame and fame is a good illustration of our general approach. An ordinary concept from everyday life, like fame, is too complex and too imprecisely defined to be quantifiable. So we search for things that we can measure, like phame, that are as close to the original concept as possible. The result is a compromise, a celebrity impersonator that we can use as our guinea pig and that we can subject to careful experimentation. As better datasets emerge that incorporate things like tabloids, magazines, and scholarly articles, phame as we measure it will become obsolete, and more sophisticated alternatives will be developed. The Wrights' wind tunnel would pale in comparison to the LENS-X turbines used today to generate Mach 30 winds for testing new spacecraft.

But for now, phame is a pretty good start. So good, in fact, that we're not going to dwell on the distinction any longer; to keep things simple, we're just going to call everything fame. Almost famous is famous enough.

Equipped with our new wind tunnel, what can we learn about the aerodynamics of a person's takeoff? And about the grim mechanics of the fall back to earth?

TREATING FAME LIKE A DISEASE

As we began to study fame using the ngram data, we quickly realized that every story was different. When we tried to identify patterns, the results seemed hard to explain and even self-contradictory. We were stuck in a bottomless pit of data.

To see why we were stuck, we need to take a trip through time to 1930, to a little town in Norway called Kristiansand. There, a local doctor named Kristian Andvord was struggling to understand the epidemic that was devastating his patients and his nation. Andvord was studying tuberculosis, which afflicted Norway to an extent we might find hard to fathom today. In the Norwegian city of Trondheim, for instance, more than 1 percent of babies born between 1887 and 1891 died of tuberculosis before reaching their first birthday. Among children between the ages of eleven and fifteen, nearly half of all deaths were attributable to the disease.

At the time, it was easy to see that something peculiar was going on. As the decades-long epidemic wore on, the average age of Norwegian tuberculosis victims was increasing. How could that be?

Andvord (or, according to an apocryphal story, a nurse working with him) had an idea. Instead of studying disease incidence over time in the entire population, he should break the population up into cohorts, groups of people who were born at roughly the same time. The advantage of this approach was that by controlling for birth year, he could do a far better job of accounting for misleading effects, such as a famine that might have only affected a single generation of children. The disadvantage was that this approach required a lot more data than could possibly be collected in the little town of Kristiansand.

Like Zipf, Andvord hit the road on a quest for data. To the great fortune of Andvord and of medical history, the Norwegian government had been meticulous in its efforts to track mortality statistics. Andvord was able to get government data covering the entire period from 1896 through 1927. He supplemented the Norway results with additional datasets from England, Wales, Denmark, and Sweden. Armed with this wealth of information, Andvord could now ask and answer the simple questions that had stymied him before. For instance, at what age were the people born in 1900 (the 1900 cohort) most likely to die of tuberculosis? What about the 1910 cohort? What about the 1920 cohort?

The answers he obtained were astonishing. It turned out that, regardless of their year of birth, disease victims were most likely to contract tuberculosis between the ages of five and fourteen or between the ages of twenty and twenty-four. Andvord's cohort analysis revealed that tuberculosis was primarily a disease of the young, and had been all along.

But if so, how could it be that if one looked at the entire population, the average age of tuberculosis victims was increasing over time? The crucial insight came when Andvord examined the total incidence of the disease—the likelihood that a member of a particular cohort would die of tuberculosis at some point in their life, young or old. As Andvord examined younger and younger cohorts, he found that the total incidence got lower and lower. Norwegians born in 1920 were less likely to contract tuberculosis in their lifetimes than Norwegians born in 1910, who were in turn less likely to contract tuberculosis than Norwegians born in 1900, and so on.

This cast the common finding about age in a different light. It wasn't that the disease was targeting increasingly old people; it was that people born earlier were more vulnerable to contracting tuberculosis throughout their whole lives. The immediate consequence of these findings was a medical bombshell: Young Norwegians were becoming progressively more resistant to tuberculosis. The epidemic was functioning as a murderous but effective mass vaccination campaign.

Though completely unexpected, Andvord's astonishing conclusions proved to be correct. This was not his only legacy. Andvord's cohort method was a revolutionary insight that has become an essential scientific tool for epidemiology and public health. Wherever massive datasets about public health are being aggregated, Andvord's ideas are likely at work. It is to Andvord (or, possibly, his nurse) that we are indebted for such knowledge as the association between high blood pressure and cardiovascular disease, the association between cigarette smoking and lung cancer, the association between blood sugar and diabetes, and tens of thousands of other associations which ensure that our every dietary decision is riddled with guilt.

Like studies of tuberculosis, studies of fame are confounded by all sorts of generation-specific effects. For instance, the invention of the Internet has dramatically influenced how people become celebrities. In our initial research, these generation-specific effects made it extremely hard to see what was going on.

Finally, we did what any good data scientist should have done in the first place. We asked ourselves, WWAD? ("What Would Andvord Do?") Suddenly, the solution became clear. We should use the cohort method. We should treat fame like a disease.

THE HALL OF FAME

At the time, we had just met Adrian Veres. A truly stellar undergraduate, Adrian knew a thing or two about immortal fame: For winning first place in the Intel International Science and Engineering Fair, he had already had a minor planet named after him, 21758 Adrianveres.

Working with Adrian, we began to create cohorts consisting of the individuals in each generation who were the most severely afflicted by fame: the Twains, the Gandhis, the Roosevelts. We chose to study people born between 1800 and 1950. Earlier, and we would be wading into parts of the dataset where our data quality was not at its very best. Later, and we would be unable to track fame across a sufficiently long period: Someone born in 1950 frequently won't get famous until the '80s or '90s, giving us only a handful of years' worth of usable data. Adrian analyzed hundreds of thousands of people, computing the frequency of mention of their full names (for example, *Mark Twain*). For each year between 1800 and 1950, he generated a list of the fifty most famous people born that year. It was particularly impressive work, considering that Adrian had just turned six on his home planet. If celebrity is a disease, Adrian's lists contained its 7,500 worst victims.

The groups were an exciting set of people who revealed the many diverse paths to fame. Take the cohort, or class, of 1871. The fifty most famous people born in 1871 included Orville Wright, our inspiration, who became famous when he learned how to fly. Ernest Rutherford became famous for his remarkable scattering experiments, which revealed the existence of the atomic nucleus. And Marcel Proust became famous for writing good books.

The class valedictorian—the most famous person born in 1871—was Cordell Hull. Never heard of him? He's much less known now, but in his heyday, Hull was a titanic figure. A United States senator, Hull eventually became the longest-serving secretary of state. His eleven years under Franklin Delano Roosevelt spanned the height of World War II. Among other things, Hull played a huge role in founding the United Nations, for which he was honored with the Nobel Peace Prize. Roosevelt himself referred to Hull as "the Father of the United Nations." The head of the class really made good.

Each and every class comprises a similar pastiche of fascinating life stories. The class of 1904 includes the Chilean poet Pablo

Neruda, the Surrealist painter Salvador Dalí, and Robert Oppenheimer, leader of the Manhattan Project, which built the first atom bomb. Its valedictorian is Deng Xiaoping, the Chinese leader. The 1899 valedictorian is Ernest Hemingway; that class includes the Argentine writer Jorge Luis Borges, the actors Fred Astaire and Humphrey Bogart, the iconic director Alfred Hitchcock, and the gangster Al Capone. An invitation to that reunion dinner would be an offer you can't refuse.

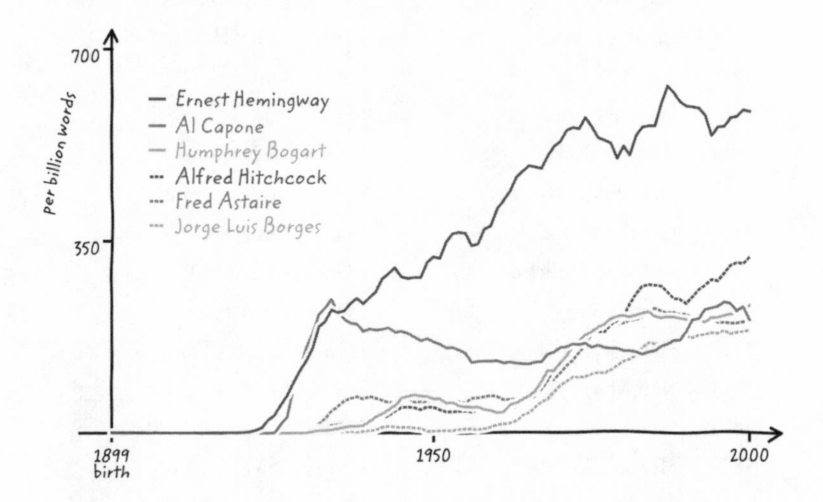

The 150 valedictorians are listed in the table that follows. See how many of the names you recognize. You can think of this as the most objective history test you'll ever take. These names don't reflect our opinion of whom you should know about, or the opinion of a teacher or professor or scholarly authority on world history. Instead, they reflect the aggregate opinion of *everyone* who has written a book in English since 1800.

1800	George Bancroft
1801	Brigham Young
1802	Victor Hugo
1803	Ralph Waldo Emerson
1804	George Sand
1805	William Lloyd Garrison
1806	John Stuart Mill
1807	Louis Agassiz
1808	Napoleon III
1809	Abraham Lincoln
1810	Leo XIII
1811	Horace Greeley
1812	Charles Dickens
1813	Henry Ward Beecher
1814	Charles Reade
1815	Anthony Trollope
1816	Russell Sage
1817	Henry David Thoreau
1818	Karl Marx
1819	George Eliot
1820	Herbert Spencer
1821	Mary Baker Eddy
1822	Matthew Arnold
1823	Goldwin Smith
1824	Stonewall Jackson
1825	Bayard Taylor
1826	Walter Bagehot
1827	Charles Eliot Norton
1828	George Meredith
1829	Carl Schurz
1830	Emily Dickinson
1831	Sitting Bull
1832	Leslie Stephen
1833	Edwin Booth
1834	William Morris
1835	Mark Twain
1836	Bret Harte
1837	Grover Cleveland
1838	John Morley
1839	Henry George
1840	Crazy Horse
1841	Edward VII
1842	Alfred Marshall
1843	Henry James
1844	Anatole France
1845	Elihu Root
1846	Buffalo Bill
1847	Ellen Terry
1848	Grant Allen
1849	Edmund Gosse
1850	Robert Louis Stevenson
1851	Oliver Lodge
1852	Brander Matthews
1853	Cecil Rhodes
1854	Oscar Wilde
1855	Josiah Royce
1856	Woodrow Wilson
1857	Pius XI
1858	Theodore Roosevelt
1859	John Dewey
1860	Jane Addams
1861	Rabindranath Tagore

1862	Edward Grey
1863	David Lloyd George
1864	Max Weber
1865	Rudyard Kipling
1866	Ramsay MacDonald
1867	Arnold Bennett
1868	William Allen White
1869	André Gide
1870	Frank Norris
1871	Cordell Hull
1872	Sri Aurobindo
1873	Al Smith
1874	Winston Churchill
1875	Thomas Mann
1876	Pius XII
1877	Isadora Duncan
1878	Carl Sandburg
1879	Albert Einstein
1880	Douglas MacArthur
1881	Pierre Teilhard de Chardin
1882	Virginia Woolf
1883	William Carlos Williams
1884	Harry Truman
1885	Ezra Pound
1886	Van Wyck Brooks
1887	Rupert Brooke
1888	John Foster Dulles
1889	Jawaharlal Nehru
1890	Ho Chi Minh
1891	Hu Shih
1892	Reinhold Niebuhr
1893	Mao Zedong
1894	Aldous Huxley
1895	George VI
1896	John Dos Passos
1897	William Faulkner
1898	Gunnar Myrdal
1899	Ernest Hemingway
1900	Adlai Stevenson
1901	Margaret Mead
1902	Talcott Parsons
1903	George Orwell
1904	Deng Xiaoping
1905	Jean-Paul Sartre
1906	Hannah Arendt
1907	Laurence Olivier
1908	Lyndon Johnson
1909	Barry Goldwater
1910	Mother Teresa
1911	Ronald Reagan
1912	Milton Friedman
1913	Richard Nixon
1914	Dylan Thomas
1915	Roland Barthes
1916	C. Wright Mills
1917	Indira Gandhi
1918	Billy Graham
1919	Daniel Bell
1920	Irving Howe
1921	Raymond Williams
1922	George McGovern
1923	Henry Kissinger

1924	Jimmy Carter	1937	Saddam Hussein
1925	Robert Kennedy	1938	Anthony Giddens
1926	Fidel Castro	1939	Lee Harvey Oswald
1927	Gabriel García Márquez	1940	John Lennon
1928	Che Guevara	1941	Bob Dylan
1929	Martin Luther King, Jr.	1942	Barbra Streisand
1930	Jacques Derrida	1943	Terry Eagleton
1931	Mikhail Gorbachev	1944	Rajiv Gandhi
1932	Sylvia Plath	1945	Daniel Ortega
1933	Susan Sontag	1946	Bill Clinton
1934	Ralph Nader	1947	Salman Rushdie
1935	Elvis Presley	1948	Clarence Thomas
1936	Carol Gilligan	1949	Nawaz Sharif

We were curious how well people would do at recognizing these, the most famous people of bygone years, so we did a completely unscientific poll. We asked a professor in the department of history at Harvard, who identified 116 of 150. A history grad student we know managed 123; a journalist, 103; a recent college grad, 73; a Russian theoretical physicist, 58; an undergraduate student in Singapore, 35.

Although people varied a great deal in terms of which names they recognized, some valedictorians were unknown to everybody, like 1868's William Allen White, an influential newspaper editor and an important progressive leader; or 1886's Van Wyck Brooks, a Pulitzer Prize–winning historian and an early biographer of Mark Twain. Remember Cordell Hull? Sadly, only the history professor did.

On some level, it's remarkable that we all don't recognize

each and every one of these names. When we study history in high school, we learn about thousands of specific personalities. But those individuals reflect a choice, a decision on the part of whoever creates the curriculum that certain figures are more important for us to know about and others less so. Dickinson, for instance, benefited from a posthumous decision by literary critics who decided that her work really mattered, despite its having made almost no impact during her lifetime. We invest the people who make such choices with a great deal of power—the power to shape our view of history. It's not immediately obvious that any person, or any small group of people, should really have that power.

On the other hand, it's clear from a quick look at this list that it, too, cannot be the basis for the historical narratives that we pass on to our children. Of the 150 valedictorians, only 16 are women; the vast majority are Caucasian men. This list has its own deep biases.

Who is at fault? For once, it's not the people who wrote down the list. Our list may have many shortcomings, but emphasis on our personal opinions is not one of them. We just crunched the numbers. Instead, the bias we observe is the collective responsibility of the true authors of the list: anyone who has ever written a book. It is the intrinsic bias of the historical record. And on some level, it must be reflected not only in our list, but in all historical research. Whether one reads books by the dozen, as a historian might, or books by the million, as we do, we're all sampling from the same massive collection. No one is immune to sampling bias. History may play favorites, but statistics do not.

Of course, the understanding that the historical record is extremely biased is an old one. What ngram data lets us do, however, is start to measure the bias, giving us a clearer picture of what we're doing wrong. If we can better remember the bias in our past, perhaps we are not condemned to repeat it.

THE GRANDEE UNIFIED THEORY

In the future, everyone will be world-famous for fifteen minutes.

—Whatshisname

Andy Warhol once made a keen observation about the fickle nature of fame. But we think he got the numbers wrong.

Let's use our Fame Hall of Fame to uncover his mistake. Viewed from up close, each of these celebrities is completely different. Some of them grew up as accomplished wunderkinds. Others are late bloomers. Some are multitalented, whereas others stick to what they do best. Some have long careers filled with one achievement after another. Others are one-hit wonders. But from a distance, these differences start to disappear, and shared features become more apparent. This is the great power of Andvord's cohort method.

When we look at the average behavior of the fifty most famous people born in 1871 (Cordell Hull's class), a single shape emerges, an overall portrait of how the class of 1871 got big. We can do the same thing for the class of 1872. Again, a single shape emerges. What's remarkable is that even though the class of 1872 consists of fifty completely different people, the shape of its average fame curve is almost exactly the same. Indeed, the shape is almost exactly the same for each and every one of the 150 classes we studied. This shape is the typical lifestyle of the extremely famous. If fame were physics, this would be the Grand Unified Theory. Or at least, it would be some sort of theory.

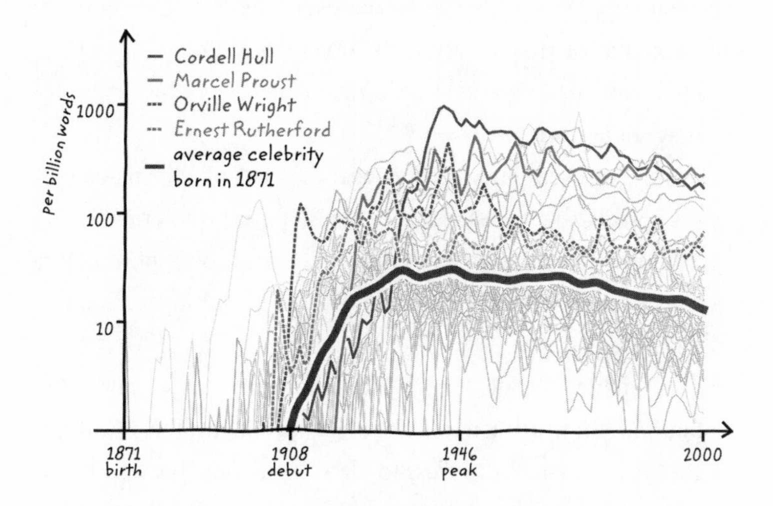

Let's take a more careful look at what's going on.

Initially, no signal is observed: For a long time, members of the class are almost never mentioned in books. That's no surprise. When twelve-year-old Orville Wright was pedaling around on a bicycle, no one was writing books about young Orville's pronouncements that one day he was going to fly.

At some point a few decades after their birth, the class members make their debut on the social scene. By debut, we mean that their average frequency is greater than one part per billion—that's the cutoff frequency for getting a word into the dictionary that we discussed in the previous chapter. By our lights, the standard for being famous is that your name deserves to be in the dictionary.

But these are no ordinary debutantes. Their arrival is not greeted by a quick flurry of interest, followed by a quicker exit.

Instead, the class of 1871, like every other class of famous people, bursts onto the scene with tremendous energy. Its members' fame rises at an extraordinary pace. Every handful of years, its average frequency doubles, skyrocketing over a period of decades. In the language of mathematics, its growth is exponential—like a viral epidemic or a viral video. On the great stage of history, theirs is a bravura performance.

Finally, at seventy-five years of age, the class of 1871 reaches its peak. Crossing that threshold, they are, in purely numerical terms, over the hill. What comes next for them must be a new experience, as these once-youthful firebrands enter a slow decline that will last for centuries.

This shape—debut, exponential growth, peak, and slow decline—is universal across all the classes we studied. But there are subtle variations from class to class, variations that can be described in terms of three parameters: their age at debut, the speed of their exponential growth, and the rate of their post-peak decline. Mathematically, a fourth parameter is needed to describe this curve: the age at which the class is over the hill. But as best we can measure, this doesn't seem to change much. All classes peak roughly three-quarters of a century after their birth.

Let's talk about the age at debut, when a class becomes so famous that half of the members are as frequently discussed as the typical word in the dictionary. For the class of 1800, this occurred at age forty-three. Not too bad, we think to ourselves—we've still got time. But the age at debut is getting younger and younger. In fact, by the mid–twentieth century, it had declined to twenty-nine years.

This fact is worth contemplating: By the time they were twenty-nine years old, half of the class of 1950 had reached dictionary-level mention frequencies in English books. Making them really, really famous.

For most of us, this is a phenomenally sobering state of affairs. For instance, when we made this discovery, JB was twenty-eight years old. Just under the wire, there was still hope for young JB, though it was clear that he had better make his move soon. Erez, however, was thirty. It was already too late.

This is particularly useful information if your goal is to be one of the most famous people of your generation. For our ambitious readers in their teens and twenties, this ought to be a subtle reminder to get cracking. Readers in their thirties should be aware that they are already running late. Readers over forty probably warrant some external guidance. We'll come to that in the next section. (Do not be dismayed. Strategies exist for winning fame well into your golden years.)

Not only do people get famous at a younger age, but their fame grows faster. For the class of 1800, it took about eight years for fame to double, allowing about four doublings between their debut at age forty-three and their peak at age seventy-five. For the class of 1950, the doubling time was much faster, only about three years.

As a result, although the shape of the curve is the same, the younger classes get *much* more famous than the older classes. As diseases go, fame is the opposite of tuberculosis. The curve looks the same for each cohort, but instead of being more resistant to fame, younger cohorts are more likely to be afflicted. The most famous people alive today are exponentially more famous than their predecessors.

To give a sense of how famous these classes can get, it's helpful to compare them to objects we encounter every day. Consider the produce aisle. At his peak, the 2-gram *Bill Clinton* was almost exactly as frequent as the word *lettuce*, twice as frequent as the word *cucumber*, and about half as frequent as the word *tomato*. He completely outclassed second-tier vegetables like *turnip* and *cauliflower*. We won't even bring up the sorry fate of *rutabaga* and *kohlrabi*.

The third parameter examines how fast fame declines after the peak. Like a radioactive element or an irregular verb, the fame of the famous has a half-life, a characteristic period of time over which it tends to decline by half. The timescale for this parameter has also been getting shorter. In 1800, this half-life was 120 years. By 1900, the half-life had dropped to 71. People are getting more famous, but they are also forgotten faster. So forget about what ol' whatshisname said: In the future, everyone will be world-famous for only 7.5 minutes.

Fortunately, the extremely famous have nothing to worry about. They should bear in mind the story of the man who, upon hearing at a conference that the sun will die out in 4.5 billion years, sighed loudly and remarked, "What a relief! I thought it was 4.5 million years." By the time the decreasing half-life of fame begins to affect them appreciably, the extremely famous will be extremely dead.

HOW TO GET FAMOUS: A GUIDE TO CHOOSING YOUR CAREER

Some of you are probably young enough that you have yet to make the great and momentous decision, "What do I want to be when I grow up?" Should you become a writer, inspiring audiences through the power of your words? A film actor, bringing characters to life with authentic displays of emotion amid simulated displays of explosion? Should you be a singer? A dancer? A teacher? A police officer? A politician? A rock star? Do you want to be the first astronaut to land on Mars, or the next Pablo Picasso? All these options are open to you.

One of the great challenges in choosing a career is the lack of solid data, some way to know what your life would be like if you picked one option or another. That's why, when you ask people what you should do with your life, their advice is always so vague.

But we're numbers guys. The loosey-goosey "Follow your bliss" type of advice that everyone else has been giving you is just not our style. Instead, we're going to present you with cold, hard statistics, quantitative data to help you make a difficult decision.

Assuming, of course, that the only thing you care about is becoming very, very famous.

We assembled focus groups composed of celebrities born between 1800 and 1920, broken down by their chosen occupation. We looked at six possible career choices: actor, writer, politician, scientist, artist, and mathematician. In each case, the twenty-five most famous members of that profession were included in our focus group. If you're considering becoming a stockbroker, a barista, or a cartoon character, then, alas, you're out of luck—we didn't have enough room in our chart.

Of course, you don't just want to know how famous you can become in each profession. Becoming really famous is no use if you're long dead or if you're too old to enjoy it. This would be like accepting a high-paying job in which the first paycheck won't arrive for a century. To make an informed decision, what you want to know is how famous you can expect to be throughout your life (assuming everything goes off without a hitch and, as expected, you manage to become one of the most famous members of your chosen vocation). So that's exactly the chart we put together for you.

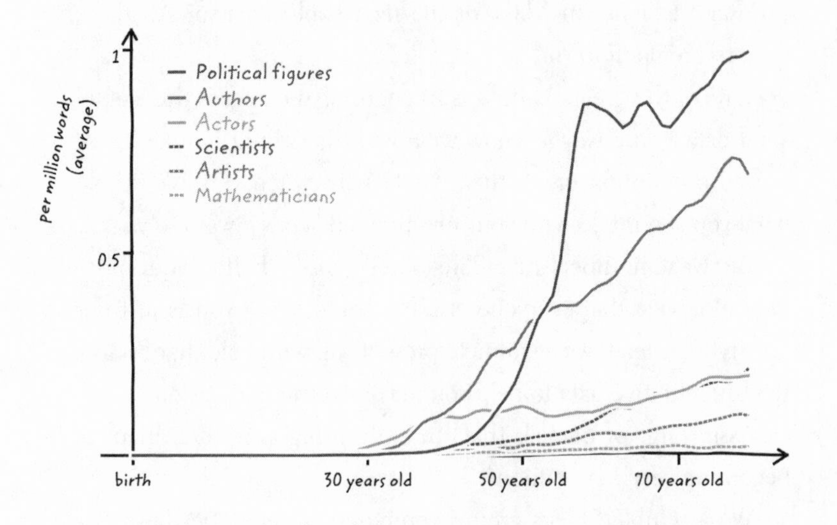

This chart will make your decision much, much easier.

If you really want to be young and famous, be an actor. Actors tend to become famous in their late twenties or early thirties, and have a lifetime to enjoy their fame. But the actors we studied lived before mass media like television could have helped propel their careers, and never got quite as famous as some of the other groups.

If you're willing to delay gratification for a little while longer, it makes more sense to become a writer. Writers tend to become famous in their late thirties, but the best writers—those who penned great classics—eventually become much more famous than the actors. This is particularly true when you look at books, since writers like to write about other writers. (Sampling bias again: the ngram equivalent of a home-field advantage.)

Contrary to what you might expect, if you're really good at delaying gratification, you should probably become a politician. Politicians tend not to be very famous until their forties, fifties, or even sixties, at which point the most famous politicians get elected president of the United States (in eleven of the twenty-five cases) or become the head of state somewhere else (an additional nine cases), and their fame quickly soars to eclipse either of the other two groups. So if you're in your fifties and not yet a household name, politics might just be your calling.

Next we took a look at scientists. The most famous scientists eventually become about as famous as the actors, but they took a lot longer to get there, achieving fame in their sixties instead of their twenties. Less fame, longer wait. It's definitely better to star on *The Big Bang Theory* than to study the big bang theory.

Still worse is drawing pictures of the big bang theory, or of anything else. The artists on our list got a raw deal. They waited just as long for fame as the scientists, but did half as well.

But if you want to become famous, the worst possible thing to do is what we did: to pursue mathematics.

You might think this is not so. After all, mathematicians are said to do their best work when they are young, after which they can presumably put their feet up and relax. For example, when he was nineteen years old, Carl Friedrich Gauss invented modular arithmetic, proved the law of quadratic reciprocity, conjectured the prime number theorem (one of the deepest and most fundamental results in all of mathematics), and discovered a profound result about the decomposition of integers into triangular numbers. This is not everything he did when he was nineteen years old; this is just what he did over a span of about three months. What a show-off.

The problem is that the public does not care what mathematicians like young Carl Friedrich are doing. By the time the mathematicians in our focus group managed to generate an appreciable fame signal, most of them were dead. Math won't make you famous. QED.

INFAMY

We know when people get famous, how fast they get famous, how quickly they'll be forgotten, and even which career choices lead them to fame. But it's impossible to conclude our discussion of fame and the ngrams without asking a very simple question: When all is said and done, who are the most famous people born in the last two centuries?

To examine the most famous people, we'll need to change our methods a little bit. The strategy that we've used so far—tracking mentions of people's full names—is great for looking at one person or a group of people over time. But when comparing different individuals, there are all kinds of peculiar effects that make full-name frequency a poor choice.

For instance, consider the following totally unsurprising fact. When referring to most people, writers tend to use that person's last name rather than write out a full name. If you see the word *Einstein*, the chance that the previous word was *Albert* is only about one in ten.

But if a person's first name and last name are both only one syllable long, people will write out their full name much more often. If you see the word *Twain*, the chance that the previous word was *Mark* is better than 50/50.

The simplest way to solve this problem is to stop tracking mentions of a person's full name and instead track mentions of their last name. An extra advantage of this is that you catch a lot more mentions, for the reasons pointed out above. The big disadvantage is that some extremely famous people, like Franklin Delano and Teddy Roosevelt, have last names that are ambiguous. Both of them account for a very large proportion of *Roosevelt* mentions, making it impossible to use our data to get accurate numbers for either one.

Another important thing to note is that our approach can't distinguish between fame and infamy. The ngram data doesn't give us enough context, enough of the words that appear before or after the name in question, to determine if a mention is positive or negative.

But alas, much as they gnaw at us, we're going to have to set those issues aside. At this stage in the game, lists like ours can be regarded only as a work in progress—at best a Wright-style wind tunnel, and certainly not a LENS-X turbine.

With that, here is a list of the ten most famous people born in the last two centuries:

1. Adolf Hitler
2. Karl Marx
3. Sigmund Freud
4. Ronald Reagan
5. Joseph Stalin
6. Vladimir Lenin
7. Dwight Eisenhower
8. Charles Dickens
9. Benito Mussolini
10. Richard Wagner

It's impossible not to be struck by the fact that Adolf Hitler, one of the most evil men in human history, tops the list. In fact, no fewer than three mass murderers appear on the list: Hitler, whose Nazi regime murdered between ten and eleven million innocent civilians and prisoners of war; Joseph Stalin, leader of the Soviet Union, whose regime killed approximately twenty million of its own citizens; and Benito Mussolini, dictator of Italy while it formed part of Hitler's axis, and architect of the Ethiopian genocide that led to three hundred thousand deaths.

Murder and fame are linked. A tragic fact about the contemporary United States is that from time to time, deranged lunatics bearing guns engage in public killing sprees. One of the many paradoxes of this terrible phenomenon is the extent to which the killer, who was a complete unknown before the event, becomes the center of a massive media storm. On the one hand, this sort of news coverage is important, because people need to be aware of what has happened. But on the other hand, the resulting attention can become a motivation for the killers. Mark David Chapman, who murdered John Lennon, said as much when he told his parole board: "I did it for attention. To, in a sense, steal John Lennon's fame and put it on myself."

Tragically, when we examine the historical record at the grandest possible scale, a similar sort of effect seems to hold true. We used ngrams to go back in time, and we generated a list of the ten most famous people for each of the last twenty decades. When we examine the list circa 1940, neither Hitler nor Stalin appears. But by 1950, after carrying out atrocities of unprecedented magnitude and cruelty, Hitler, Stalin, and Mussolini jump to numbers 1, 2, and 5, respectively. In contrast, Abraham Lincoln, perhaps the greatest and most morally courageous of American presidents, never appears above number 5.

As we've seen, exploring fame using ngrams can be intriguing, perplexing, and even fun. But darkness, too, lurks among the ngrams, and no secret darker than this: Nothing creates fame more efficiently than acts of extreme evil. We live in a world in which the surest route to fame is killing people, and we owe it to one another to think about what that means.

Does it have to be this way? Here, too, the ngrams can provide us with a hint. Because the person who preceded Hitler at the top of the fame list, holding the number one spot from 1880 to 1940, was not a mass murderer. He was a writer, a social critic, a "genial and loving humorist," and by most accounts a good man. He may even have given us "Merry Christmas!"

Charles Dickens. Peace and war. It was the best of times, it was the worst of times.

One giant leapfrog for mankind

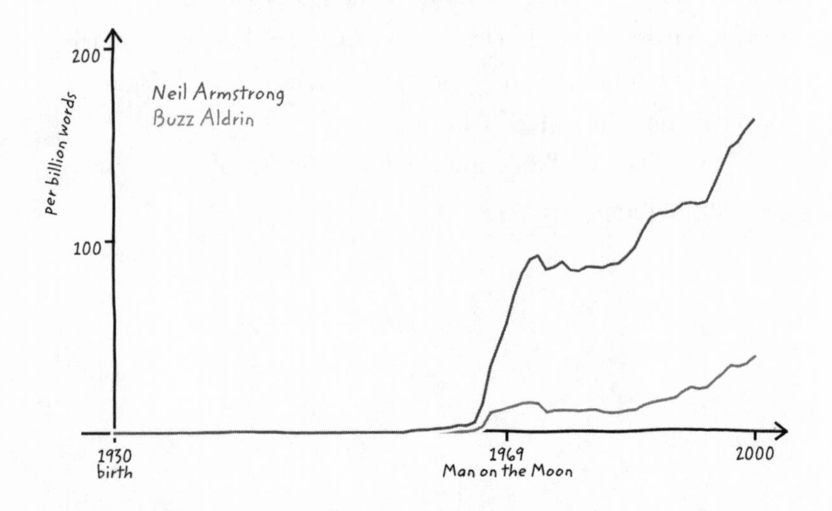

The USSR's launch of the Sputnik satellite in 1957 captured the world's imagination and heralded the space race. That race was won by the United States on July 21, 1969, when two Americans landed on the moon and went for a stroll.

More specifically, the space race was won by Neil Armstrong, who traveled 239,000 miles to become the first human to walk on the surface of a distant world. You've probably heard of him.

You're much less likely to have heard of another American hero, Buzz Aldrin. Aldrin walked on the moon, too, thereby fulfilling a dream mankind has probably shared for tens of thousands of years. And he also did it on July 21, 1969. But he wasn't first: Aldrin took his small step nineteen minutes and one one-hundredth of a second after Armstrong. As a result, he's about five times less famous.

The moral: If you're planning to do something legendary, do it before your twenty-minute coffee break.

5

THE SOUND OF SILENCE

Dort wo man Bücher verbrennt, verbrennt man auch am Ende Menschen.

Where they burn books, they will, in the end, burn people.

—Heinrich Heine (1797–1856), German
Jewish poet blacklisted by the Nazis in 1933

The millions of voices reflected in books tell a long and fascinating story about our culture and our history. But not everyone's voice is recorded on our bookshelves. And sometimes the silence of the missing voices can drown out everything else.

One of the people whose voices our culture nearly missed out on was Helen Keller. Born in 1880, she was left deaf and blind by an illness she contracted when she was only nineteen months old. Keller came of age in an era when such disabilities made it nearly impossible for someone to become educated. But she persevered. As the first deaf-blind person to earn a bachelor's degree, Keller eventually grew to be an influential author, a political activist, and an eloquent advocate for the needs of the disabled. In the process,

Keller became a hero to millions, a symbol of the triumph of the human spirit over profound adversity.

And yet at one of the darkest moments in human history, Keller had to confront an attempt to silence her voice—and the voice of a legion of others—once again.

In 1933, the Nazis began taking over Germany, aiming to control its government, its people, and even its culture. One manifestation of this movement was the suppression of books believed by the authorities to reflect an "un-German spirit." Urged on by Nazi leaders, mobs of students forcibly removed such books from libraries and bookstores and set them aflame in book burnings that erupted all across Germany. Included among the blacklisted authors was Helen Keller.

Keller's response, an open letter published on the front page of the *New York Times* and many other newspapers, was and remains a timeless cri de coeur:

May 9, 1933

To the student body of Germany:

History has taught you nothing if you think you can kill ideas. Tyrants have tried to do that often before, and the ideas have risen up in their might and destroyed them.

You can burn my books and the books of the best minds in Europe, but the ideas in them have seeped through a million channels and will continue to quicken other minds. I gave all the royalties of my books for all time to the German soldiers blinded in the World War with no thought in my heart but love and compassion for the German people.

I acknowledge the grievous complications that have led to

your intolerance; all the more do I deplore the injustice and unwisdom of passing on to unborn generations the stigma of your deeds.

Do not imagine that your barbarities to the Jews are unknown here. God sleepeth not, and He will visit His judgment upon you. Better were it for you to have a mill-stone hung around your neck and sink into the sea than to be hated and despised of all men.

Helen Keller

Keller's impassioned argument that "history has taught you nothing if you think you can kill ideas" struck a chord the world over. It touched off an international furor, eventually leading the Nazi propaganda machine to frame the book burnings as unofficial "spontaneous acts by the German Students Association."

Though Keller carried the day in the court of world opinion, was she actually right? Is it really impossible to kill an idea? Our quest to answer this question will force us to tackle the dark side of human expression: the world of censorship, of suppression, and of infamy. To get a glimpse of this dark reality, there are few better windows than the life of the most famous of all window-wrights, the artist Marc Chagall.

A STAINED-GLASS WINDOW

"Go and find a book in the library, idiot; choose any picture you like; and just copy it."

This advice on how to draw, from a schoolmate, launched the extraordinary artistic career of Móyshe Shagal, transforming the son of a herring trader from Vitebsk, Belarus, into "the quintessential Jewish artist of the twentieth century," Marc Chagall.

A pioneer of the modernist movement, Chagall was one of the leading artists of the mid–twentieth century. He is famous, above all, for his stained-glass windows. Fusions of color, glass, and light, his *Jerusalem Windows* are an Israeli national landmark—they have even appeared on the nation's postage stamps. Chagall's windows also grace the United Nations and illuminate cathedrals throughout Europe. "When Matisse dies," Pablo Picasso once said, "Chagall will be the only painter left who understands what color really is."

Like many of the famous people discussed in the previous chapter, Chagall became prominent at a young age. After the Russian Revolution of 1917, when he was only thirty years old, Chagall was offered the position of commissar for the visual arts over all of Russia. But war and famine were taking their toll on Russian life. Soon, despite being one of the nation's most famous young artists, Chagall headed west to Paris.

When he arrived in Paris in 1923, Chagall was not as well known, and he had to work hard to reestablish himself. He was exquisitely aware of the impact that his choice to emigrate had made on his fame and reputation. Chagall confides as much in a letter to Pavel Ettinger, a collector and commentator back in Russia:

March 10, 1924

I'm afraid that my "image" is little by little . . . fading. . . . It is no wonder. I have been here for quite a while, in the homeland of painting. What shall I say about myself. I could say a lot, but I have to be brief. Gradually, they are beginning to notice me in France. . . .

Needing to be brief, Chagall sums up his recent experience by saying that "they are beginning to notice me in France," while at the same time expressing fear that his image is "fading" back home. This concern, the centerpiece of an intimate letter between longtime correspondents, is remarkably quantitative: How often are people thinking, talking, and writing about Chagall?

Of course, Chagall lacked any precise way of measuring how famous he was and in which direction his fame was going. But at least to the extent that his fame led to mentions in books, it's easy for us to examine.

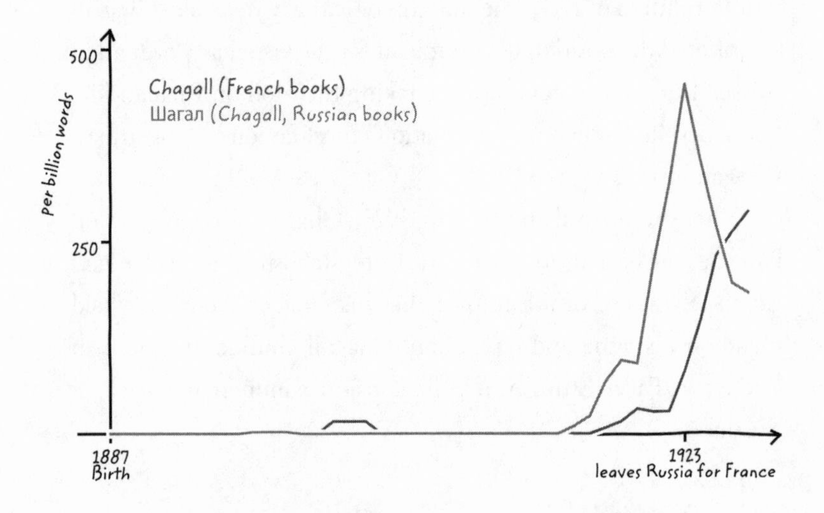

Chagall's assessment was dead-on. We can readily see the effects of his choice to emigrate, which were already quite pronounced by the time of his letter to Ettinger.

But Chagall's prominence would soon be affected by events well beyond his control. On the other bank of the Rhine, a brown army was massing. Avant-garde artists, like Chagall, would soon be dubbed "un-German." And Chagall's situation was even more precarious: He was a Jew.

DEGENERATE ART

In the 1920s, Germany was a haven for the arts. Dada, Bauhaus, Expressionism, and Cubism had all taken root there. But Adolf Hitler strongly objected to these styles. He was a failed artist with conservative tastes. Moreover, the freewheeling nature of these new movements was contrary to his plan of using culture as a form of social control.

To justify the draconian control of German culture that Hitler hoped to exert, the Reich relied extensively on the theories of a turn-of-the-century critic named Max Nordau. Nordau claimed that many aspects of modern culture, such as avant-garde art, were a product of hitherto-unrecognized mental diseases, such as dysfunctions of the visual cortex. On this basis, the Nazis argued that it was necessary to rid German culture of such influences, which they labeled Jewish, notwithstanding the fact that Nordau himself was Jewish, and an important Zionist leader to boot. In September 1933, Hitler allowed Joseph Goebbels, Reich minister of propaganda, to create the Reichskulturkammer (Reich Culture Chamber). His mission: to carry out Hitler's plans for purifying German culture.

Under Goebbels, the Reichskulturkammer became by far the

most important institution in German artistic life. Goebbels announced, "In the future, only those who are members of a chamber are allowed to be productive in our cultural life. Membership is open only to those who fulfill the entrance condition." Among other things, membership required showing a certificate of Aryan ancestry and demonstrating a willingness to go along with the ideology of the Reichskulturkammer. Thus, Goebbels could safely conclude, "In this way all unwanted and damaging elements have been excluded." The Nazis were not content merely to hamstring artists by means of Kafkaesque membership requirements. In June 1937, Goebbels appointed Adolf Ziegler, one of Hitler's favorite painters, to head a new commission within the Reichskulturkammer. Its task was to confiscate art that the Nazis considered degenerate from collections, public and private, throughout the country.

As a Jewish surrealist expressionist, Chagall was right in the crosshairs, and his works soon began to disappear from Germany. At the same time, thousands of other "degenerate" pieces were taken, including works by many of the most famous modern artists in the world today—Georges Braque, Paul Gauguin, Wassily Kandinsky, Henri Matisse, Piet Mondrian, and Pablo Picasso. Some of the confiscated pieces were destroyed, some were kept by Nazi leaders, and some were hidden away in places like the Altaussee salt mine. The effect on the art world is hard to underestimate. (When Edvard Munch's *The Scream* was put on display at the Museum of Modern Art in New York in 2012, the heirs of a German Jewish banker who once owned the piece insisted that MoMA should include a note pointing out that their father had been forced to sell the painting after the Nazis rose to power.)

THE MOST POPULAR ART EXHIBIT OF ALL TIME

Confiscating avant-garde art and prohibiting those who produced it from making more was not enough. Goebbels and Ziegler didn't just want to eliminate modern art in Germany, they wanted to discredit it. To this end, they set about creating two side-by-side art exhibitions in Munich. One exhibition highlighted artists who had the approval of the regime. The other featured works that Ziegler and his cronies had been busily confiscating. In his 1937 speech inaugurating the exhibits, Ziegler issued an invitation: "German *Volk*, come and judge for yourselves!"

The first exhibition, called the *Große Deutsche Kunstausstellung* (Great German Art Exhibition), was one of the most lavish art exhibitions in modern history. In fact, it was not just art that was on exhibit: The show inaugurated the Haus der Kunst (House of Art), a monumental new museum building that was a showpiece of Nazi architecture. On display were numerous works by Nazi-approved artists, such as Arno Breker, who sculpted physically flawless nudes in the neoclassical style.

The second exhibition, titled *Entartete Kunst* (Degenerate Art), featured many of the most famous works that Ziegler had confiscated. Pieces by Chagall, Kandinsky, Max Ernst, Otto Dix, Max Beckmann, Paul Klee, and László Moholy-Nagy were on display. But the pieces were not given the same treatment as those of the *Große Deutsche Kunstausstellung.*

This exhibit did not take place in a monumental new museum. Instead, the works were crammed into a smaller space on the second floor of a building that had once housed the German Institute for Archaeology. It was accessible only by a narrow stairwell. The pieces themselves were crowded, poorly hung, and often unframed. Works were frequently labeled with the price a museum paid to acquire them. Because many had been bought during the period of German hyperinflation in the 1920s, the numbers were particularly outlandish.

The exhibit was largely disorganized, except for sections dedicated to works that the Nazis thought demeaned religion or German military and family life. The walls were covered with graffiti-like slogans, such as "Deliberate Sabotage of National Defense," "The Ideal—Cretin and Whore," "Nature as Seen by Sick Minds," "An Insult to German Womanhood," and "The Jewish Longing for the Wilderness Reveals Itself—in Germany the Negro Becomes the Racial Ideal of a Degenerate Art." Of the 110 artists whose works were on display, only six were Jewish, and their pieces were placed in a separate room, titled "Jewish, All Too Jewish." Nevertheless, an undercurrent of the exhibit was that all of modern art was a "Jewish-Bolshevist" conspiracy against German values.

In short, *Entartete Kunst* was not designed to be an exhibition in the ordinary sense of that word. It was, instead, the exhibition as subversive government-funded polemic. It was a propaganda piece whose goal was to undermine modern art, to present it as morally bankrupt, crassly commercial, and a waste of taxpayer funds.

And it was a huge blockbuster, attracting more than 2 million visitors in its first four months alone, or nearly 17,000 people a day. It attracted five times as many visitors as the exhibit at the Haus der Kunst. These numbers were and remain unheard-of for an art exhibition.

To give a sense of how well attended the exhibit was, consider that in 2011, the best-attended art exhibit in the world, the Centro Cultural Banco do Brasil's *Magical World of Escher*, attracted 9,677 people per day, little more than half the traffic of *Entartete Kunst*. In 2010, New York's Museum of Modern Art put on a major exhibition, *Abstract Expressionist New York*, whose subject matter overlapped somewhat with *Entartete Kunst* in that it was an exhibition of modern artists from the region. This exhibition, too, was one of the biggest of the year, drawing 1.1 million people over seven months, or about 5,600 people per day—still, just a fraction of *Entartete Kunst*.

The fact that the exhibition was popular is not merely a statistic. The massive crowds amplified the experience, becoming a part of the display. Here's how one visitor described it:

> I felt an overwhelming sense of claustrophobia. The large number of people pushing and ridiculing and proclaiming their dislike for the works of art created the impression of a staged performance intended to provoke an atmosphere of aggressiveness and anger. Over and over again, people read aloud the purchase prices and laughed, shook their heads, or demanded "their" money back.

Thus, *Entartete Kunst* was a hybrid of visual and performance art, displaying modern artwork in a tasteless and misleading way in order to incite public anger and scorn—all of which created the individual visitor's encounter. Soon, the smash hit began traveling from city to city, carrying its derisive message across Germany. All in all, between 5 and 10 percent of Germans paid a visit. Tragically, *Entartete Kunst* was the most popular art exhibition of all time.

After *Entartete Kunst*, it was effectively impossible to be a modern artist in Germany. Beckmann, Ernst, Klee, and several other artists fled the country. Those who remained were forbidden to create art. Emil Nolde, facing such a ban, secretly continued painting in watercolor so that the smell of paint would not give him away. Ernst Ludwig Kirchner finished the job the Nazis had begun: He committed suicide.

And what of Chagall? Even as his name was being rapidly effaced from German culture, Chagall, living in France, was initially safe from physical violence. But when France fell in 1940, Chagall realized that his life was in danger. Using forged visas, his family left for the United States.

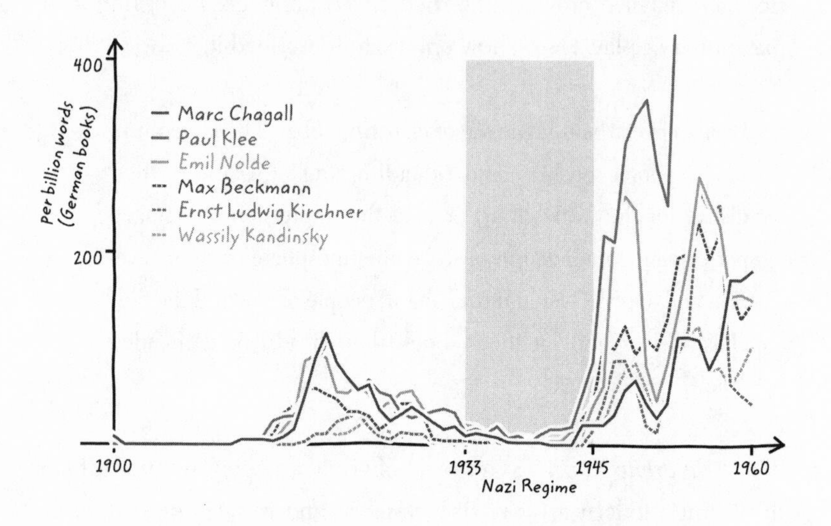

These ngrams, computed from books published in the German language, make the effects of Nazi suppression on Chagall and his contemporaries crystal clear. Between 1936 and 1943,

Marc Chagall's full name appears only once in our German book records. The Nazis did not manage to kill Chagall. But they found a way to erase him.

BOOK BURNINGS

The Nazi regime's manipulation of German culture extended far beyond modern art, shaping every aspect of German thought. Any concept that the regime deemed unsuitable was a target. In this campaign against ideas, books were an inevitable and early battleground. Less than ten weeks after Hitler was sworn in as chancellor, the battle was joined.

Nazi influence had so deeply penetrated German society that the opening salvo in this battle did not come directly from the government. In April 1933, the principal student union in Germany, called the Deutsche Studentenschaft, initiated a nationwide campaign to cleanse German culture of undesirable ideas. Within days, in a conscious attempt to echo Martin Luther, the students put up posters all over Germany, listing "12 Theses Against the Un-German Spirit." Here is thesis number 7:

> We want to regard the Jew as alien and we want to respect the traditions of the *Volk* [the German people]. Therefore, we demand of the censor: Jewish writings are to be published in Hebrew. If they appear in German, they must be identified as translations. Strongest actions against the abuse of German script. German script is only available to Germans. The un-German spirit is to be eradicated from public libraries.

In the thrall of the Nazi movement, the Deutsche Studentenschaft had come to believe that the roots of German problems lay, among other places, in libraries, in the form of works reflecting the "un-German spirit." But the students had a problem: As we know, it's hard to read all the books in the library. How would they know which ones reflected the "un-German spirit"?

For this, they needed Wolfgang Herrmann, a librarian who had joined the Nazi Party in 1931. Obscure and often unemployed, Herrmann had spent years combing the stacks to compile lists of books that he thought were a bad moral influence. Herrmann was extremely meticulous in this personal obsession, creating separate lists for all sorts of authors, including politicians, literary writers, philosophers, and historians.

None of his efforts would have mattered much, except that, as Hitler rose to power, Herrmann's profile rose, too. Named to a "purification committee" tasked with overhauling Berlin's libraries, Herrmann was suddenly in a position to begin his own campaign against what he called Germany's "literary bordellos." The Deutsche Studentenschaft turned to Herrmann to ask him to share his meticulously curated lists with its campaign. These he willingly provided. Within months, the once-obscure librarian had an army at his disposal and Germany's libraries in his sights.

On May 10, 1933, the initial campaign reached its climax: the *Säuberung* (cleansing). Outfitted with torches and Herrmann's lists, students took to the streets of most of Germany's university towns, raiding bookstores, lending libraries, and schools, consigning tens of thousands of books to the flames. In Berlin, they were led by Goebbels himself, who announced that "the era of extreme

Jewish intellectualism is now at an end. . . . The future German man will not just be a man of books, but a man of character." By the end of May, there had been book burnings all over Germany. Five hundred tons of books had been confiscated by the Gestapo. The burned books included works by Karl Marx, F. Scott Fitzgerald, Albert Einstein, H. G. Wells, Heinrich Heine, and of course Helen Keller.

Yet even the May book burnings were only the beginning of a protracted attack by the Nazis on Germany's books. Herrmann kept revising his lists, and they swelled from about five hundred authors in 1933 to thousands of writers by 1938, becoming the core of a continually expanding blacklist supported by the regime. This sustained attack had a devastating impact. Margaret Stieg Dalton, a librarian and a historian of libraries, estimates that 69 percent of the books present prior to the regime in the public library in Essen, a Nazi industrial center, had been removed by 1938. These included many of the most widely circulated books. In a world without the Internet, the impact of removing so much information from the public sphere can hardly be imagined.

Although it's hard to envision the world that the Nazis created, in which many of the ideas that are most important to us today had been effaced from the national discourse, we can still get statistical insight into the efficacy of their censorship campaign by using ngrams. The chart below shows the fame of authors listed on Herrmann's various blacklists. For comparison, we include a list of Nazis as well.

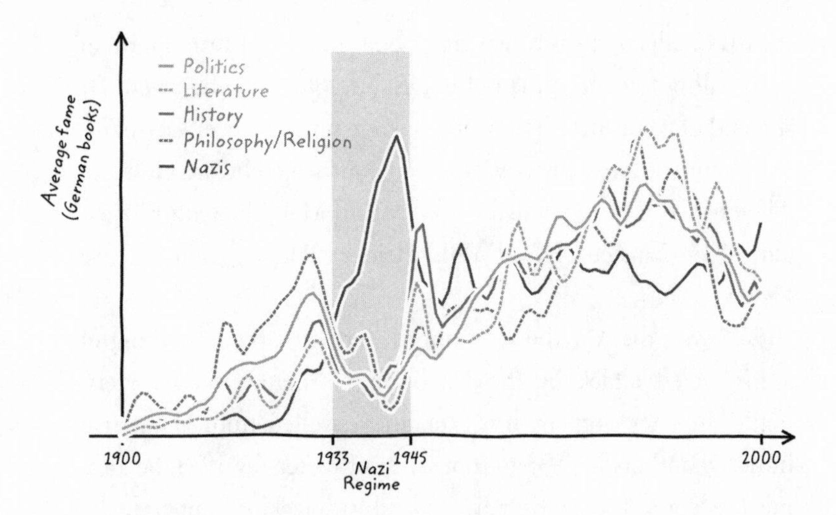

The contrast between the fame of the blacklisted intellectuals and that of people linked to the Nazi regime could not be more obvious. It renders the efficacy of Nazi suppression terrifyingly apparent.

An additional observation can be made. Perhaps surprisingly, Herrmann's campaign was not equally effective in all disciplines. For instance, the fame of authors of philosophy and religion books included on his blacklist declined fourfold during the Third Reich. The fame of authors writing about politics declined by half: less than that of the philosophers, but still pronounced. In contrast, his blacklist of historians had a more limited effect; the decline was only about 10 percent. Using ngrams, we can perceive the contours of the Nazi campaign against ideas more keenly than ever before.

WHAT THEY DON'T WANT YOU TO KNOW: A WORLD TOUR

The Nazi regime is without a doubt the best-documented case of large-scale political and cultural suppression. But although it is an extreme example, it is hardly the only one. Like a powerful searchlight, big data can reveal instances of censorship all over the world. Some of them are closer than we might like to think.

A few years after he guided the Russian Revolution that established the Union of Soviet Socialist Republics, Lenin suffered a stroke that compromised his ability to lead. A struggle for power immediately broke out. Leon Trotsky, who led the Bolsheviks along with Lenin, had been expected to succeed him. But three heroes of the revolution—Joseph Stalin, Grigory Zinoviev, and Lev Kamenev—formed a political alliance to undermine Trotsky. The *troika*'s strategy succeeded brilliantly, leading to the official denunciation of Trotsky at the XIII Conference of the Communist Party and their ascendance in his place. Once Trotsky had been neutralized, Stalin turned on his co-conspirators. By 1925, the *troika* had dissolved, and Stalin was the sole leader of the USSR.

But Stalin was not satisfied with a simple promotion. In his quest for absolute power, he began a systematic campaign to suppress every potential rival, getting rid of long-standing enemies and recent friends with equal dispatch. Figures like Zinoviev and Kamenev were isolated, expelled from the Party, put on trial, and, in 1936, executed during what is known today as the Great Purge. Already exiled in Mexico, Trotsky was nonetheless condemned to death in absentia during those same trials. His days were numbered: In 1940, Stalin sent the assassin Ramón Mercader to exe-

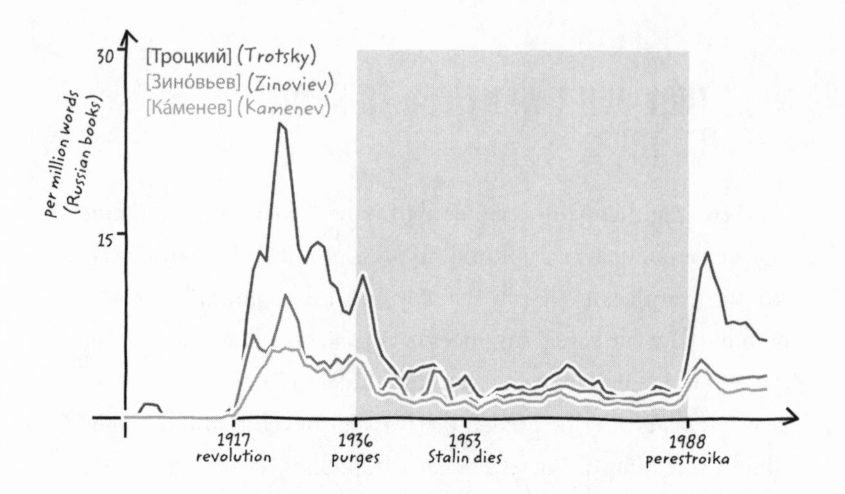

cute the court's judgment. Trotsky, hero of the revolution, died in Mexico of an ax blow to the head.

Yet even this story doesn't fully capture the impact Stalin had on his rivals. His goal was not merely to kill them. He wanted to wipe out any record of their contributions, to erase them from the memory of their countrymen, leaving himself, alone, as the central hero of the revolution. And by and large, Stalin succeeded.

For nearly half a century after their executions, the contributions of Trotsky, Zinoviev, and Kamenev, along with countless others, were minimized and ignored. As the ngrams reveal, the fame of all three drops precipitously after the Great Purge. Neither Stalin's death nor the public repudiation of the Great Purge by Nikita Khrushchev in 1956 managed to restore them to their rightful place in history. Their reputations would, eventually, be partially rehabilitated. But this took generations: We don't see the ngrams rebound until the *perestroika* (restructuring) and *glasnost* (openness) reforms ushered in by Mikhail Gorbachev in the late '80s.

Stalin was not the only person who feared these old Bolshevik revolutionaries and their dangerous ideas. In post–World War II America, anxiety about communism was on the rise. Were there communists in the United States? If so, where were they, and what were they up to? To ensure an adequate investigation, the House of Representatives established a special standing committee in 1945: the House Un-American Activities Committee.

Fearing that the film industry could become a clandestine source of foreign propaganda, one focus area for the committee was the influence of communists on Hollywood. In its 1947 hearings, the committee began by listening to the testimony of friendly witnesses, Hollywood personalities whose patriotism the congressmen did not question. Several of them, including Walt Disney and Ronald Reagan (at the time, he was only president of the Screen Actors Guild), spoke of a grave communist threat to their industry. Soon the committee turned to unfriendly witnesses who were suspected of communist ties, hoping that they would reveal what they knew and that they would name names. Under pressure, most agreed to testify. But ten refused: Alvah Bessie, Herbert Biberman, Lester Cole, Edward Dmytryk, Ring Lardner, Jr., John Howard Lawson, Albert Maltz, Samuel Ornitz, Adrian Scott, and Dalton Trumbo. Many of them had been extremely successful in their trade, even winning Academy Awards. Today, they are collectively known as the Hollywood Ten.

On account of their refusal to testify, the Hollywood Ten were cited for contempt of Congress. Worse still, forty-eight important Hollywood producers (including such figures as Samuel Goldwyn and Louis B. Mayer) weighed in, eager to bolster their anti-communist creds. The producers issued a statement announcing that not one of the Hollywood Ten would be allowed to work for

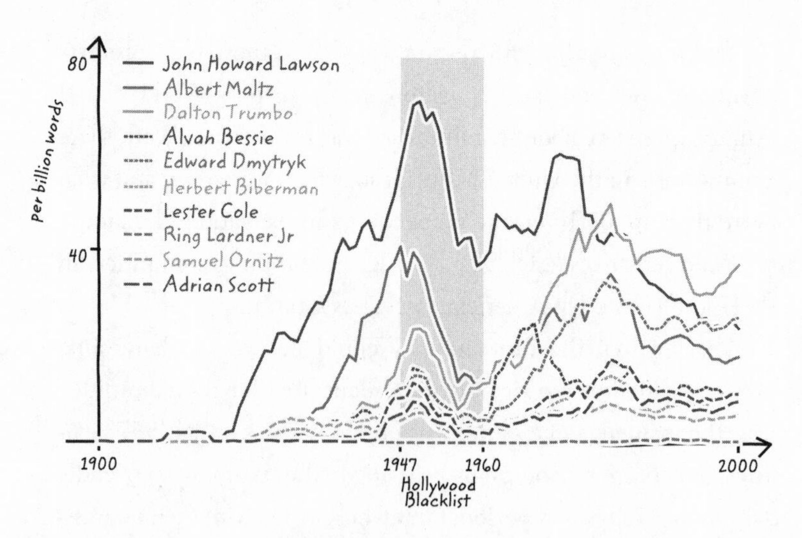

their studios "until such time as he is acquitted or has purged himself of contempt and declares under oath that he is not a communist."

With those words, the producers established a blacklist that prevented the Hollywood Ten, and later many others, from finding work in the United States. No member of the Hollywood Ten was credited in a movie produced by the major studios for over a decade. The impact on their lives and careers was immediate and devastating.

It was only after the downfall of Senator Joseph McCarthy in the mid-'50s that the power of the House Un-American Activities Committee began to wane. (Although his goals were often similar, it is important to note that McCarthy, a senator, did not play a role in the House initiative.) Former president Harry Truman put an exclamation point on this reversal with his remark in 1959 that the Un-American Activities Committee was the "most un-American thing in the country today." Stripped of public sympathy, the blacklist was primed for collapse. Finally, in 1960, the blacklist was violated, when Dalton Trumbo was credited as the screenwriter for the aptly named film *Exodus*. Hollywood's exiles had returned to the promised land.

Our history is so full of suppression that it's easy to get caught up talking about one example after another. But suppression is also happening today—perhaps more than ever before. One of the best examples is the legacy of Beijing's Tiananmen Square.

Two particularly notorious incidents have taken place in Tiananmen Square in recent memory.

In 1976, China's ruling Gang of Four cracked down on protests and public mourning in Tiananmen Square. Spurred by the death of venerated premier Zhou Enlai, about ten thousand people had come together. Although the square was cleared by force, no lives are believed to have been lost. The 1976 incident leaves a huge fingerprint in the Chinese ngram record, with a massive spike in mentions of Tiananmen (天安门).

But the far more infamous event—in the eyes of the West—was the Tiananmen Square massacre of 1989. Prompted this time by the demise of an important official, pro-reform general secretary Hu Yaobang, mourning students took to the square. Again, the public display of sorrow became a protest, in which as many as one million people are said to have participated. In response, the government declared martial law and sent three hundred thousand troops into the capital. On June 4, 1989, the troops reached the square and cleared it in an extremely violent crackdown. The number of deaths—still unknown today—is believed to have been in the thousands.

By all rights, the 1989 Tiananmen Square massacre ought to be the rallying cry of all Chinese dissidents, a flash point and a fixture in Chinese culture.

But it isn't.

After the massacre, Chinese government officials sprang into action, carrying out a campaign of censorship and information suppression that was remarkable for its speed and efficacy. Within a year, more than 10 percent of China's newspapers had been shut down, along with numerous publishing houses. To this day, all print media describing the massacre is required to be consistent with the government's account. Digital media is also monitored as part of China's extensive Internet censorship campaign, often referred to as the "Great Firewall of China." Those who search the Internet for *Tiananmen Square* see a carefully sanitized list of results. (From 2006 to 2010, Google agreed to participate in China's blockade, although it has since ended its cooperation.) As a result, many young people in China today know little, if anything, about the events of June 4, 1989. When quizzed, undergraduates at Beijing University appeared not even to recognize the "tank man" image—iconic elsewhere—showing a defiant Tiananmen Square protester standing in front of a column of Chinese tanks.

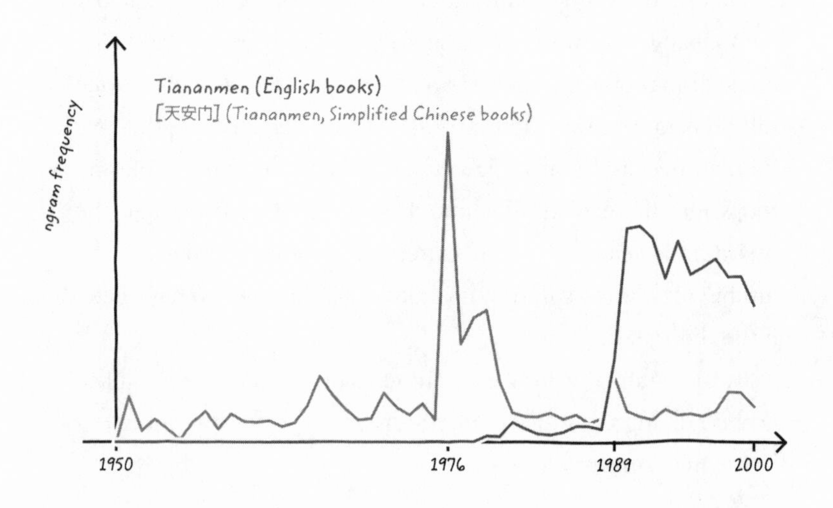

In the West, mentions of Tiananmen soar after the 1989 massacre. In China, there is a transient blip of interest—hardly approaching even 1976 levels—after which things go back to normal.

The Tiananmen Square massacre is one of the central events of contemporary Chinese history. But nobody there ever discusses it, at least not in print. Many may not even know about it. This heartbreaking chart is a testament to the brutal efficiency of censorship in contemporary China.

CAN WE DETECT CENSORSHIP AUTOMATICALLY?

No matter where they take place, censorship and suppression often leave a characteristic mark: the sudden disappearance of particular words and phrases. The statistical signature of this lexical vacancy is often so strong that we can use numbers—big data—to help figure out what is being suppressed.

Let's go back to Nazi Germany to see how this works. Our goal will be to look for people whose fame shows a Chagall-like drop during the Third Reich, 1933 to 1945. We can measure the size of this drop by comparing a person's fame during the Third Reich to the person's fame before and after. If his frequency of mention is one in ten million in the '20s and the '50s, but goes down to one in one hundred million during the Nazi regime, that's a tenfold drop. It suggests that the person was probably being censored or suppressed in some way. On the other hand, if the frequency goes up to one in a million during the Nazi era, a tenfold increase, then the person was particularly famous during the regime, and may have been benefiting from government propaganda. In this way, we can take any name and assign it a suppression score, reflecting the magnitude of the drop or of the increase. This in turn helps us figure out who was being suppressed by the surrounding society.

We applied this automated detector to thousands of names of famous people who were alive during the Second World War, and made two charts. The first chart shows the suppression scores we get for English. Most of the scores are close to one: no drop, no increase. Fewer than 1 percent of people have a score bigger than five in either direction. This chart is nothing special: The results for English are typical, and closely resemble what we see in almost all languages during almost all periods of time.

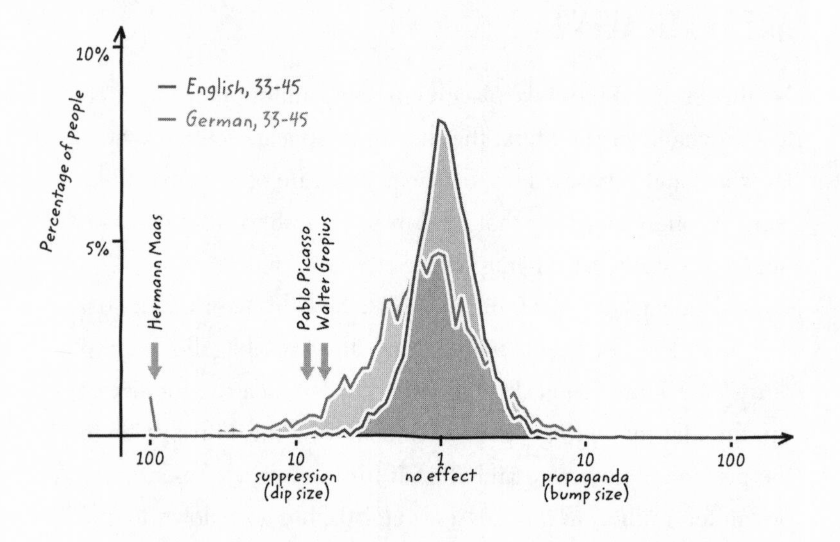

The second chart shows the results for the German language during the Nazi regime. It looks completely different. First, it isn't centered on one, but a little to the left. Most people were being suppressed, at least somewhat, by the regime; a majority suffer significant drops in fame. But it's not just that the center has moved.

The distribution is also much wider, containing far more extreme values. A few of these are on the right, where we expect to see the beneficiaries of government propaganda. But most are on the far left: More than 10 percent of the people on our list suffer a fivefold or more drop in fame.

The names on the left include Picasso. They include Walter Gropius, founder of the Bauhaus movement in art, architecture, and design. And if you go as far left as you can, you find the name of Hermann Maas, a Protestant minister who publicly denounced the Nazis and helped Jews obtain exit visas to flee Germany. For his efforts, the Reich made him the target of a personal campaign. We're certainly not the first to notice Maas' extraordinary heroism: In 1964, Yad Vashem, Israel's Holocaust memorial, recognized Maas as one of the Righteous Among the Nations.

After we made this chart, we asked a scholar from Yad Vashem to make her own personal determination, using only the tools of an ordinary historian, about which names would appear at which end of the curve. We didn't give her access to our data or to our results, and we didn't even tell her why we were asking. All she got from us was the list of names. Nevertheless, her answers agreed with ours the vast majority of the time.

So our statistical censorship-detection technique gives results that are qualitatively similar to those of a traditional historian using traditional methods. But unlike traditional methods, our analysis can be done almost instantaneously, by a computer.

Automated methodologies like this one hold a great deal of potential for our day-to-day lives. We all want to be able to identify the effects of censorship, suppression, and even just ordinary bias on the information that we consume every day. Today, censorship watchdog organizations try to help by carefully reading media in a region of interest and on a topic of interest and highlighting the omissions they find. But as more and more information is produced, it is becoming impossible to read everything, or even a significant slice of everything. We need alternatives. Big data is a powerful one.

Interestingly, Wikipedia has recently begun to take advantage of this big data approach to bias detection. There has been a long-standing discussion of an anti-female bias in Wikipedia, perhaps due to the fact that most Wikipedia editors are male. That discussion has relied primarily on anecdotal evidence. But new efforts are bringing statistical methods and ngram data into this dialogue. The goal is to clearly identify problematic trends and articles so that the shortcomings may be addressed.

In the future, such methods won't be limited to Web sites staffed by, for the most part, volunteers acting in good faith. They'll also serve to keep governments honest, and their people, and ideas, free.

SEEPING THROUGH A MILLION CHANNELS

In only a few short years, the Nazis went a very long way toward wiping out a great many ideas. They didn't like modern art, so they made works of art disappear, making an exception only for the demeaning presentations at *Entartete Kunst*. Modern artists like Chagall were driven from Europe, forced into retirement, or killed. The movement all but disappeared from Germany.

So what should we make of Keller's notion that "history has taught you nothing if you think you can kill ideas"?

On the one hand, the ideas have survived—we're talking about them right now. But on the other, it's a bit facile to pretend that this is how things had to work out. Hitler lost the war. If history had turned out differently, perhaps his campaign against ideas might have turned out differently, too.

Yet any discussion of censorship would be incomplete if it didn't touch on the unintended consequences of the very tactics used by oppressive regimes. Imagine that you were a young artist living in Germany who, despite extraordinary social pressure, remained interested in modern art. If so, you would probably be attracted to the *Entartete Kunst* exhibit, where many of the works of your heroes were on display. You could imagine it as a classroom of sorts—very large and very rowdy, but a master class, nonetheless.

This actually happened. In 1936, Charlotte Salomon managed to get admitted to the Berlin Academy of Fine Art, where she was the only Jewish student. She even won a prize there, although it was later retracted "on racial grounds." Salomon was very interested in modern art, and when the *Entartete Kunst* exhibition came to town, it was an extraordinary opportunity for her. After all, the Nazi regime had just collected many of the world's most important works of modern art and placed them conveniently at her doorstep. Better yet, they were available to be seen for months on end—so long as she managed to ignore the jeering throngs.

Salomon was deeply inspired by the works in the *Entartete Kunst* exhibition and learned a great deal from them. She later deployed many of the techniques of modern art to create one of the most remarkable autobiographies of the twentieth century. Salomon's mother, aunt, and grandmother had all committed suicide. In her memoir—told in the third person, a dark fairy tale about a girl named Charlotte—her doppelgänger struggles through a heartrending decision: "Whether to take her own life or undertake something wildly unusual."

The book reveals her struggle to live and to study art in the shadow of the Third Reich. Remarkably, the tale is told through the medium of 769 paintings. By the end of the work—which she titled *Life? or Theatre?*—Salomon has answered the question, concluding that an extremely unusual life would be preferable to no life at all. But alas, under the Nazi regime, it was not up to her: In 1943, Salomon died, pregnant, at Auschwitz.

Yet her work did not die with her. *Life? or Theatre?* was eventually returned to her father and stepmother, who had spent the war in hiding in Holland. Almost immediately, it was recognized as extraordinary. It has been called "the pictorial counterpart of Anne Frank's diary."

Perhaps the ideas of modern art did not rise up in their might, as Keller suggested, to destroy the Nazis. But Keller was at least partly correct. Despite the brutal Nazi efforts to suppress modern art—prohibiting it, confiscating it, mocking it, and murdering its practitioners—the ideas of modern art could not be killed. They would indeed "seep through a million channels," through unpredictable avenues like Salomon's visit to *Entartete Kunst*. And though Salomon herself was killed, her works did eventually "quicken other minds." Her testimony—the testimony of a modern artist, steeped in the great masters of modern art and testifying in the language of modern art—outlived the Nazi regime, and played a role in ensuring that the Nazis became the most "hated and despised of all men."

Chagall and Salomon—the teacher and the student—never met in person. But many years after Salomon's death, Chagall had the opportunity to see her work at an art festival. He was deeply moved. Chagall handled the works "so tenderly. He was very touched by them and said—good, they were good."

POSTSCRIPT

After the Nazis invaded Hungary in 1944, they began killing the country's Jews. Each day, more than ten thousand Hungarian Jews were taken by train to the Auschwitz death camp. To escape, Erez's grandfather, grandmother, father, and aunt went into hiding. Yet every morning, his grandfather emerged from their hiding place to pray, donning a pair of tefillin containing passages from the Hebrew Scriptures. He did so despite the fact that, had he been caught reading the texts of the Jewish liturgy, he would have risked paying the ultimate price.

As we were writing this chapter, Erez's father—the last of the four—passed away. He left Erez a treasured parcel: his own father's tefillin, worn each and every day of the war. They had been carefully preserved: Each letter of the century-old parchment was intact.

A million channels, indeed.

Two rights make another right

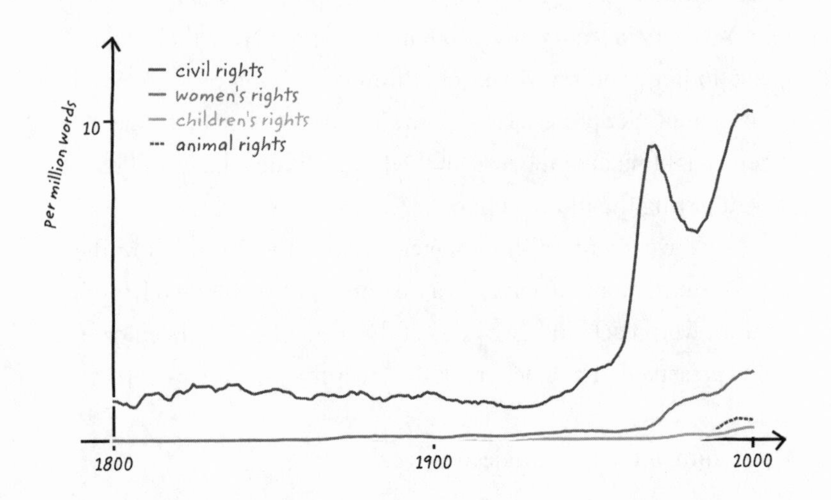

Like species, ideas can reproduce and become popular. Like species, ideas can also mutate. One example is the notion of *rights*.

One example is the notion of *rights*. This idea has a long history that traces as far back as the Roman Empire, with its concept of *ius civitatis*, the rights of the individual citizen. Energized by the theories of philosophers like John Locke (1632–1704), the concept of fundamental rights began to form the bedrock of many legal systems in the seventeenth and eighteenth centuries, through innovations like the English Bill of Rights (1689), the American Bill of Rights (1789), and the French Declaration of the Rights of Man and of the Citizen (also 1789). In the United States, the idea of *civil rights* came to refer primarily to the rights of blacks, who became a test case for how the new nation would handle racial minorities.

Encouraged by the achievements of the *civil rights* movement, other groups jumped on this ethical bandwagon. *Women's rights* first exhibits a signal after the Civil War in the 1860s, and picks up speed during the *civil rights* movement a century later. In recent decades, *children's rights* and *animal rights* have become more widespread. Today, two wrongs still don't make a right. But fortunately, too many wrongs do make a rights movement.

6

THE PERSISTENCE OF MEMORY

Before we move on, we want to tell you about one last movement to get rid of ideas.

This one differed greatly from the censorship efforts we described in the previous chapter. It was not led by a government. No blood was spilled, although, in a famous showdown, one of the principals of the movement did threaten a dissenter with a fireplace poker. And it did not begin in Germany, but across the border, in Austria, in the 1920s.

There, a group of philosophers known as the Vienna Circle had become fed up with human language, which was, in their estimation, a dreadful mess. The approach that the Vienna Circle espoused, often referred to as logical positivism, held that the only statements that make sense are those statements that can be empirically verified, that the only words that are meaningful are those that can be measured. The rest led to "inhibiting prejudices," and we'd be better off without 'em. As you might imagine, this put quite a lot of words on the chopping block. Is *love* measurable?

Can you empirically verify that something is *right* or *moral*? No, said the circle, you can't, and because these words refer to things that can't be measured, they don't belong in our language at all.

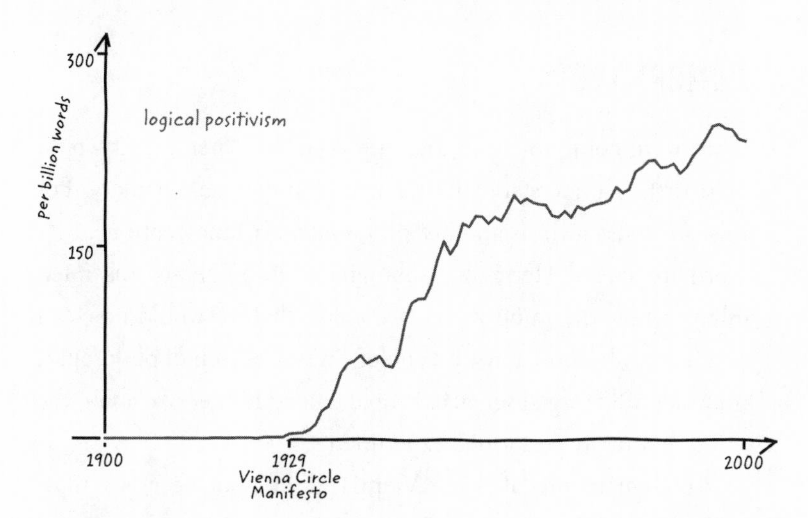

One of the circle's favorite examples was the word *Volksgeist*, "spirit of the people." *Volksgeist* was supposed to refer to a nation's collective consciousness and memory, to what the nation was like and what it had on its mind. *Volksgeist* was exactly the sort of imprecise, unmeasurable concept that irritated the circle, so the group highlighted the term in its 1929 manifesto, hoping to banish it from language altogether.

The Vienna Circle's idea-animus was less a matter of political censorship and more a matter of its philosophical attitude toward the boundaries of science.

At the time, the circle may have been right. Ideas like collective memory have long stood outside the purview of scientific investigation. But with ngrams at our disposal, probing a concept like collective memory seems less improbable. Can we measure collective memory, the same way that we might test the memory of a single person?

MEMORY TEST

If we were going to try to measure collective memory, it would help first to understand the science of individual memory. For this, we must turn to another philosopher, a nineteenth-century German named Hermann Ebbinghaus. Ebbinghaus was interested in how the mind works, a domain that we would now call psychology. In those days, psychology was a branch of philosophy, not yet a full-fledged science. People tended to theorize about the mind, but rarely performed experiments.

Ebbinghaus predated the Vienna Circle, but he was sympathetic to the idea that experience, measurement, and empirical confirmation were the foundations of human knowledge. He was not extreme enough in his beliefs to consign most of the concepts of psychology, unmeasured and perhaps unmeasurable, to the lexical scrap heap. Instead he thought that the study of the mind needed to become more empirical. As a proof of principle, he set out to do something that was, at the time, unthinkable: to investigate his own personal memory using purely experimental methods.

He immediately faced a problem that resembled the problem we faced when studying fame. Memory was a vague concept. Ebbinghaus needed to sharpen his focus by replacing the vast, ambiguous terrain of memory with a small number of well-defined, observable processes. He settled on two: how fast we learn, and how fast we forget.

Once he narrowed his scope, Ebbinghaus still faced significant challenges. Experiments benefit immensely from an isolated, controlled environment. Human memory doesn't lend itself to that. Every piece of information in our mind is embedded in a network of concepts. We associate it with related facts, ideas, people, emotions, places, times, and events. These complex relationships have a very significant effect on recall. As a result, it's very hard to study our ability to remember a particular fact in isolation. We've already seen how, by banding together, irregular verbs like *burn/burnt*, *learn/learnt*, *spell/spelt*, and *spill/spilt* can survive for centuries. These sorts of memory effects are not the exception; they are the rule.

To get around this problem, Ebbinghaus came up with an elegant solution. He realized that most associations have to do with either the sound or the meaning of what one is trying to memorize. In order to minimize unwanted associations, he decided to memorize random nonsense: a synthetic vocabulary consisting of 2,300 meaningless syllables that he had devised himself. Each syllable was just a trio of letters, consonant-vowel-consonant, like CUV and KEF. He carefully made sure that none of the syllables sounded too much like a word. This cold new world had no room for LUV, no time for a HUG, and no place for meaning.

To measure learning, Ebbinghaus would draw random nonsense syllables from his vocabulary, chaining these random syllables together into lists. He would then measure how long it took him to memorize those lists, reciting each syllable with no errors. To measure forgetting, Ebbinghaus added another step to the procedure. After learning a list, he would wait for a fixed period of time, and then see how much of the list he still remembered.

The prospect of memorizing long strings of random syllables

day after day probably didn't appeal to many potential test subjects, but Ebbinghaus did have disproportionate influence on one volunteer: himself. So in 1878, Ebbinghaus began to study memory, using himself as the only test subject.

For more than two years, he stuck to a painfully strict schedule, dedicating long stretches of time each day to memorizing random nonsense syllables. He learned list after list, using a highly regimented system, repeating the syllables at a constant rhythm dictated by the ticking of a mechanical watch. He systematically explored many combinations of variables—the length of the list, the time of day, the amount of time he spent memorizing, the position of particular syllables in the list, the time interval between repetitions, and so on. Ebbinghaus was one of the most dedicated researchers in the annals of psychology.

And nature rewarded him for it with an ensemble of spectacular discoveries. For instance, Ebbinghaus discovered that, as the lists got long, the impact on learning time of even a single additional syllable could be disproportionately large. This relationship, between the number of items memorized and time, is today called the learning curve, and when people talk about a "steep learning curve," they're referring—whether they know it or not—to Ebbinghaus. Ebbinghaus also made important discoveries about forgetting. He noticed that after only twenty minutes, he typically forgot nearly half of the list. But forgetting seemed to slow down; even a month later, he could still remember about a fifth of the list. The relationship Ebbinghaus discovered between forgetting and time is called the "forgetting curve."

Taken together, the learning curve, the forgetting curve, and the procedures used to discover them laid the groundwork for the modern scientific study of human memory. The notion of a nonsense syllabary was such an effective innovation that it remains a central method in psycholinguistics to this day. Indeed, Ebbinghaus' work was a foundational moment for modern psychology as a whole. And, of course, his personal dedication to the study itself was extraordinary. William James, a founding father of psychology, later remarked on Ebbinghaus' extraordinary dedication, lauding him for his "heroism in the pursuit of true averages." James also called the memory study "the single most brilliant investigation in the history of experimental psychology."

At first, collective memory seemed like a hard thing to probe, but Ebbinghaus' story gave us cause for optimism. The things he had managed to measure—learning and forgetting—have close analogues in human culture, which are very apparent in the ngrams.

UNFORGETTABLE

Some things are hard to forget. More than a decade after two planes barreled into New York's World Trade Center, the memory of that day still haunts Americans. Ten years later, Jon Lee Anderson, a staff writer at the *New Yorker,* recalled his experience:

> With a sense of rapidly growing horror, I saw the second plane hit and realized that it was a terrorist attack and, when the buildings collapsed, that the attack was akin to a second Pearl Harbor. I knew that my country would soon be at war.

This is not an infrequent comparison, and rightly so. Roughly sixty years before the morning of 9/11, Americans woke up to the first attack on their home soil in decades. On the morning of December 7, 1941, hundreds of Japanese planes swarmed the Hawaiian base of Pearl Harbor, dropping bombs and torpedoes, leaving smoke and fire and death in their wake. In little more than an hour, the Japanese destroyed numerous airplanes and ships, crippling the Pacific Fleet. The attack left more than 2,400 Americans dead and more than 1,000 wounded. The shocking news changed the course of history, thrusting the United States off the sidelines and into World War II.

But important though it was at the time, the better part of a century has passed since Pearl Harbor, and the attack no longer figures frequently in daily conversation. It may be hard to imagine right now, but 9/11 is on the same course.

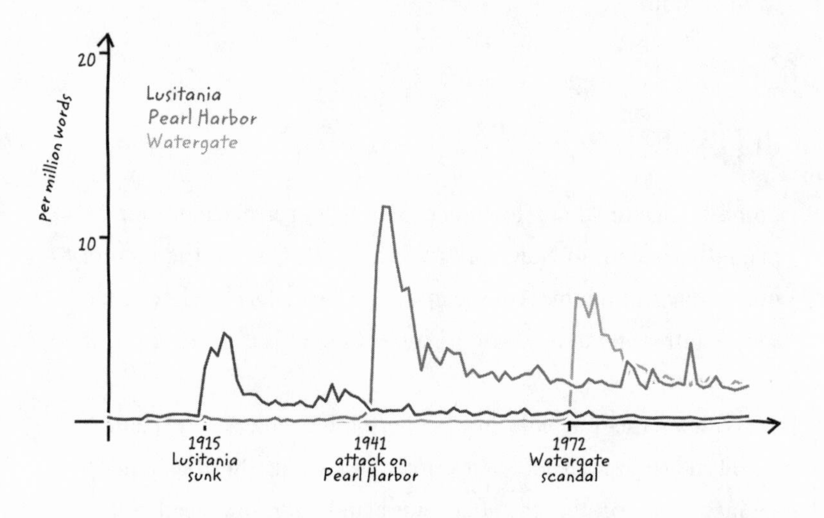

How does that happen? How does our collective memory wipe out even the most painful events?

A MEMORY BY ANY OTHER NAME

To probe this, we face an Ebbinghaus-like problem: Forgetting is so idiosyncratic, so dependent on which ideas we associate with which other ideas, that it's hard to do a good experiment.

Consider the sinking of the ocean liner *Lusitania,* which brought about America's entry into World War I. In the decades that follow the tragedy, it starts to be forgotten, much as we might expect, but it recovers, briefly, ahead of World War II, likely because concerns about a second world war brought the events surrounding the first one back to the fore. This sort of memory-by-association effect is a major problem: It's impossible to account for and impossible to predict.

An equally tricky problem is that, over time, changing associations cause people to remember the same events in different ways, using different words. Again, the world wars furnish an excellent example. *World War I* was originally called *the Great War,* as it was the deadliest war in the history of Western civilization up to that point. But as *World War II* began to erupt at the end of the '30s, the term *the Great War* quickly disappeared, replaced by the term *World War I.* Crucially, it's not that people stopped thinking about *the Great War.* Those events were still deeply lodged in the collective memory. But people thought about the war differently, in the broader context of both conflicts, so they used different language. Again, this sort of effect is impossible to account for and impossible to predict.

If we're going to measure forgetting, we'll need to emulate Ebbinghaus, minimizing the effects of all these associations by using a carefully chosen vocabulary.

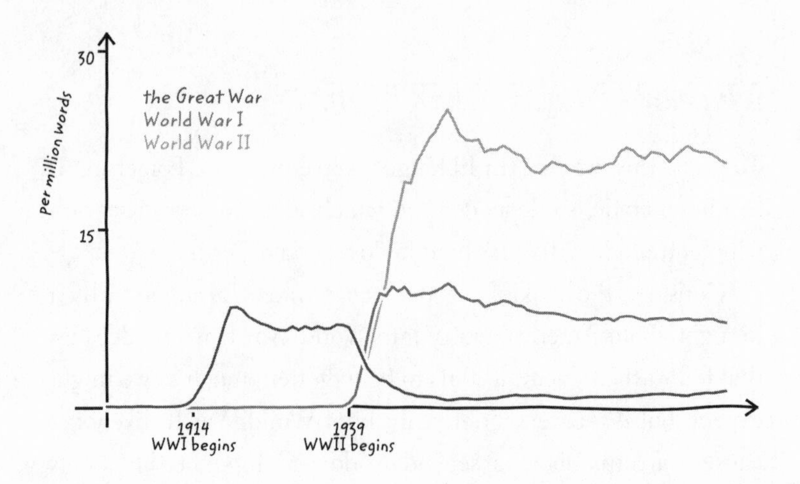

In order to do just that, we decided to try to probe collective memory using only numerals that correspond to years, like *1816* and *1952*. By seeing how often people talk about a year, we can get a sense of how present the events of that year are in their minds. No year is at a particular disadvantage, and no year is so strongly associated with any other year that it interferes too much with this crude approach.

But wait, you say. What if the sentence that the number came from was "1876 oysters on the half-shell and a glass of Picpoul, please"? In that case, the number is a reference to the number of oysters being ordered.

It turns out that this is not a significant problem. First, it would be very strange to order *1876* oysters, especially with only one glass of wine. But more important, it's very strange to order, request, or record *1876* of anything. The number *1876* comes up incredibly infrequently—except when people are referring to the year 1876. Even titles of books, like George Orwell's *1984*, and movies, like

Stanley Kubrick's *2001: A Space Odyssey*, make a negligible contribution to the overall totals for their respective numerals.

The 201 numbers between 1800 and 2000 can play the role, in the study of collective forgetting, that the synthetic vocabulary of Ebbinghaus played in the study of individual memory. What do these numbers teach us?

THE FORGETTING CURVE

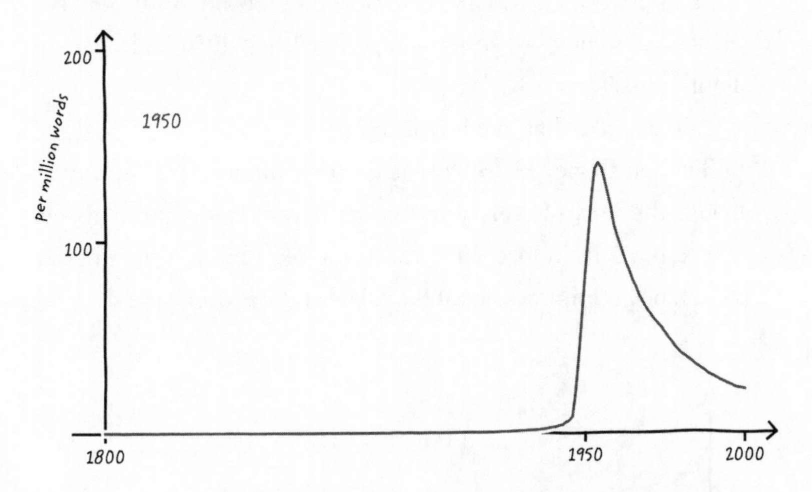

Let us tell you the story of the year 1950.

For most of human history, no one gave a damn about 1950. No one cared about it in 1700, no one thought about it in 1800, no one was concerned about it in 1900. This apathy persisted through the '20s and '30s and into the '40s.

But starting in the early '40s, there was a bit of a buzz: People realized that 1950 was going to happen, and that it could be big.

Still, nothing got people interested in 1950 like the year 1950 itself.

Suddenly, everyone was obsessed with 1950. They couldn't stop talking about all the things they did in 1950, all the things they were planning to do in 1950, all the dreams they hoped might come to pass in 1950.

In fact, 1950 was so fascinating that for several years thereafter, people felt the need to debrief. They just kept talking about all the amazing things that had happened in 1950, all through '51, '52, and '53. Finally, in 1954, someone—probably someone very fashion-conscious—woke up and realized that 1950 had become slightly passé.

And just like that, the bubble burst.

Though tragic, 1950's story is far from unique. The history of 1950 is the story of every year that we have on record: Boy meets year X, boy falls in love with year X, boy leaves year X for a newer model, boy reminisces about his X less and less over time.

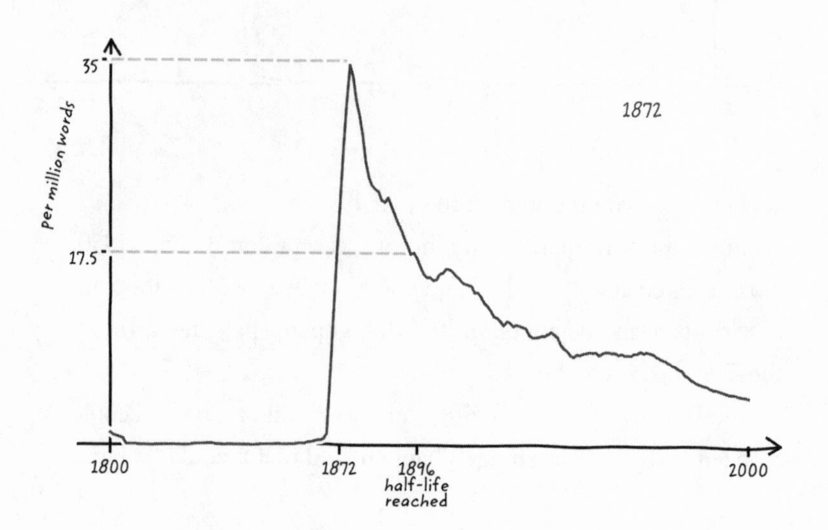

We can make these handy charts showing this same process for every year. The tale of love and loss that we just described is evident on each and every chart, but that's no surprise. Other features of these charts are more unexpected.

One such feature is the overall shape of these forgetting curves. The forgetting process seems to be composed of two regimes: Interest in a given year drops quickly in the first few decades and much more slowly thereafter. It's a striking similarity between collective and individual recall: Society has both a short-term and a long-term memory.

We can also ask very quantitative questions. For instance, let's consider society's short-term memory. We might wonder: How fast does the bubble burst? How quickly do people lose interest in a year once it has ended?

A simple approach to that question is to see how long it takes for the frequency of a year to decline to half of its peak value: the half-life of collective forgetting. This value varies substantially from year to year. The frequency of *1872* declined to half its peak value in 1896, a lag of twenty-four years. In contrast, *1973* dropped to half its peak value by 1983, after only a decade.

The speedier decline of *1973* is a symptom of a general phenomenon: As time passes, the half-life of collective forgetting gets shorter and shorter. What this observation suggests is that our society's attitude toward the past is changing. We are losing interest in past events faster and faster.

What caused that change? We don't know. For now, all we have are the naked correlations: what we uncover when we look at collective memory through the digital lens of our new scope. It may be some time before we figure out the underlying mechanisms.

This is the frontier of science. There is no map, lots of guesswork, and plenty of blind alleys, but there's no place better.

OUT WITH THE OLD, IN WITH THE NEW

Of course, our collective consciousness does more than just forget. If we're to understand collective memory, we also need to probe the other side of the coin. How does new information enter a society?

We think of our current era as the information age, a period marked by the sensational speed with which information can be passed from one person or place to another. But we lose sight of how quickly raw information could travel in centuries past, using mechanisms whose potential we no longer fully appreciate. In seventeenth- and eighteenth-century London, for example, what we now call snail mail used to arrive as often as fifteen times a day. Letters mailed in the morning would arrive within four hours. It's not as quick as today's e-mail, to be sure, but not as slow as today's snail mail, either. (By the nineteenth century, Londoners could ship parcels around the city, at speeds of up to twenty-five miles per hour, via a now-abandoned network of pressurized tubes.) For centuries, humans have had ways to ensure that big news travels fast.

Books are not one of those ways. Books are an important way to get information out, but most books are relatively large undertakings that take years to write and publish. They are much too slow for breaking news.

Often, that's not a problem. Because collective forgetting—at least, of the most important things—is relatively slow, its progress over years, decades, and centuries is easy to chart using book-derived ngrams.

But many of the things entering the collective consciousness enter quickly, in days, weeks, months, or at most a handful of years. The ngram *1872* only took a single year to make the transition from near obscurity to peak popularity. *Pearl Harbor* took but a day. Trouble is, book ngrams just aren't very useful when we're trying to measure such fast processes. You need a high-speed shutter to take a picture of a fastball.

If we are going to use our ngrams to learn about learning, we need to look at something that moves more slowly than big news.

EUREKA

Erez's wife, Aviva, began exploring one approach to collective learning that seemed particularly promising: the study of inventions. Successful inventions are the very epitome of collective learn-

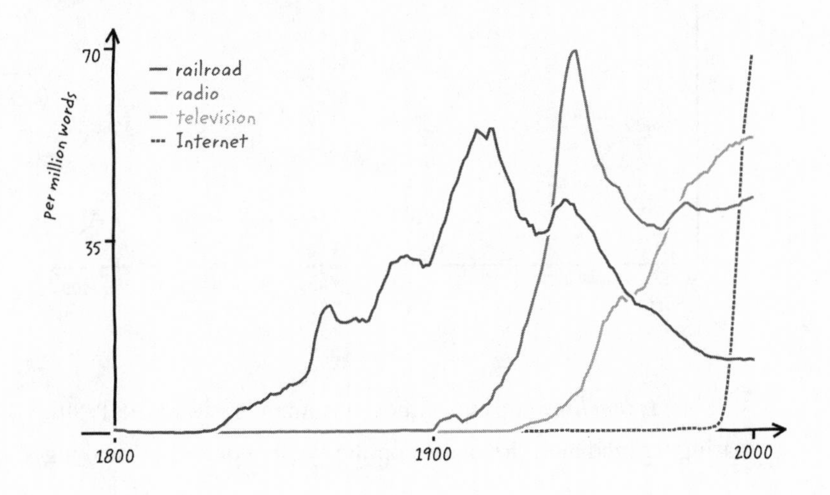

ing. They reflect society's ability to generate new knowledge about the world and to assimilate these advances in science and engineering to overcome relevant day-to-day challenges. For those very reasons, inventions take much longer to spread than ordinary news.

The crucial difference is that an invention is not just pure information that can be easily communicated via e-mail or pony. The engineering know-how to create an embodiment of the invention, the technical skill to use it, the economic model to motivate its sale and distribution, and the infrastructure to help spread it all are necessary for a society to fully embrace a new technological idea. Unlike word of a newsworthy event, it can take decades for news of an invention to propagate.

These lengthy timescales should be easy to explore using the ngrams. A great example is the fax machine.

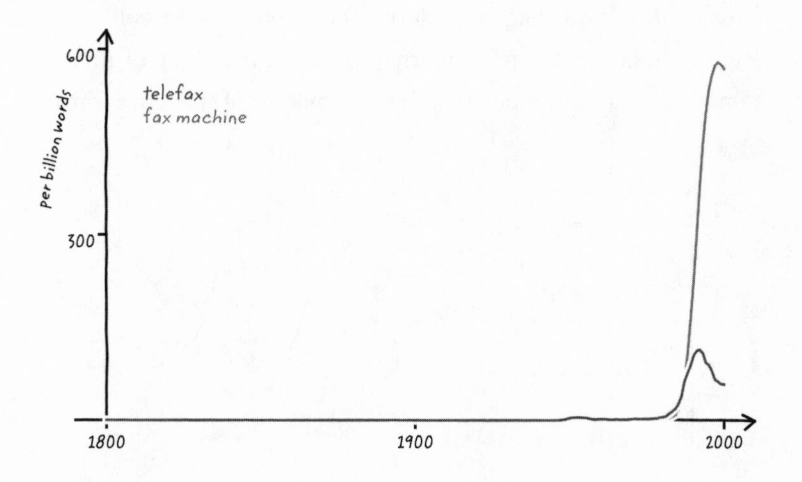

The *fax machine* pops up, almost instantaneously, in the 1980s, soaring immediately to peak popularity. It looks like breaking news. Judging by this ngram, when would you guess that the fax machine was invented?

The '80s, right? Nope. The '70s? Nope. '60s? '50s? '40s?

You got it: The fax machine was invented in the '40s. But not the 1940s. The first patent for the fax machine was awarded to Scottish inventor Alexander Bain in 1843. By 1865, a commercial service for what was then called a *telefax* had been established between Paris and Lyon—before the invention of the telephone. One of the cutting-edge technologies of the 1980s was partly backed by Napoleon III, emperor of France. Big news travels fast—but big ideas don't.

PATENT CAVEAT

If we want to examine how long inventions take to spread, we need to start with a long list of technologies and figure out when each of them was invented.

You would think that this is an easy thing to do. Governments have been awarding patents on new inventions for centuries, giving inventors the exclusive right to profit from their creations. As Abraham Lincoln—the only U.S. president to hold a patent—put it, "The patent system added the fuel of interest to the fire of genius." Patent laws encourage inventors to disclose their new technologies as soon as possible. So all we need to do to figure out when something was invented is to find the patent that was issued, and check the date.

But this, too, is easier said than done.

Consider the telephone. In the United States, the invention of the telephone is credited to Alexander Graham Bell.

On March 10, 1876, Bell wrote the following entry in his notebook:

> I then shouted into M [the mouthpiece] the following sentence: "Mr. Watson—come here—I want to see you." To my delight he came and declared that he had heard and understood what I said.

Bell later commercialized this technology, creating a series of companies whose various offshoots and offspring still dominate the telecommunications industry. To Americans, Bell is a technology hero who laid many of the foundations that enable our present information age.

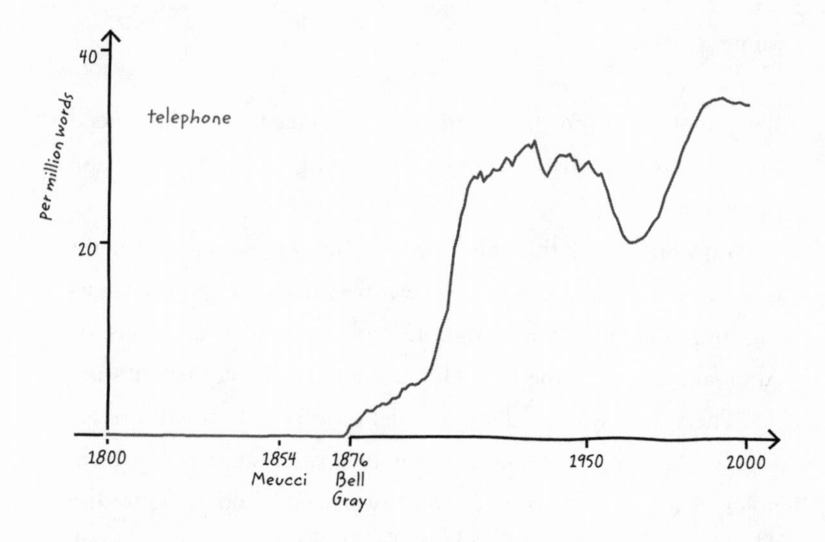

But that's not how they tell it in Italy. To Italians, the inventor of the telephone is Antonio Meucci. This Italian-American claimed to have invented a *telettrofono* around 1854 and kept improving on his design until 1870, when he managed to propagate his voice through wire for a distance of more than a mile. Watson, working with Bell in 1876, was only in the next room.

And what about Elisha Gray? Gray founded the Western Electric Manufacturing Company in 1872, which supplied telegraphic equipment to Western Union. Fiddling with this technology, Gray ended up inventing the variable-resistance microphone. This device made it possible to encode multitonal sounds, like human voices, for transmission over a wire. In effect, Gray invented the telephone, too.

The list of great minds that may or may not have invented the telephone reads like a who's who of late-nineteenth-century innovators. Many of them have patents in their name describing their contributions. Meucci filed a patent caveat—a sort of provisional patent—in 1871, calling his technology a speaking telegraph. But does that mean Meucci deserves the credit? Oddly, he let this claim expire some years later, and it never became a full patent. Furthermore, it's not totally clear that Meucci ever built exactly what he claimed to have built. On February 14, 1876, nearly five years after Meucci's filing, Gray's lawyer entered the patent office in Washington, D.C., to file a patent caveat for the invention of the telephone. That suggests that credit ought to go to Gray. But earlier that day, Bell's lawyer had entered the same office. He had filed a patent for—you guessed it—the invention of the telephone.

Don't even get us started on the lightbulb.

147 BLIND DATES

Unambiguously determining when something was invented is impossible. We needed to compromise. One option is to try to go through inventions, like *telephone*, one by one, and take our best guess based on the evidence. But that was dangerous. Perhaps our own biases, conscious or subconscious, would influence the results. Instead, Aviva did the smartest thing she could: She gave up and used Wikipedia.

Wikipedia lists dates for numerous major inventions. We know that some of them are not the best possible dates. But because they aren't our dates, we can be sure that they don't reflect our biases, and that they are unlikely to be systematically skewed in a way that will undermine our experiment. Sometimes blind dates are better.

Aviva checked each date to make sure it was plausible—that at least one of the most relevant patents was filed at that time, and that—per ngrams—the technology was not in wide use before that date, by any name (e.g., neither as *fax machine* nor as *telefax*). If the date wasn't plausible, she struck the invention from our little registry. Anything else, she kept.

What she was left with was a list of 147 big ideas and their 147 birthdays. This list includes all sorts of cool gadgets. One is the *typewriter*, patented in 1843 by Charles Thurber. (Interestingly, he thought of it as a particularly useful aid for "the blind . . . and the nervous.") Another upstanding entry is the *brassiere*, patented in 1913 by Sigmund Lindauer. The list includes molecules (*morphine* and *thiamine*), materials (*Pyrex* and *Bakelite*), methods of transportation (*helicopter* and *escalator*), ways of blowing things up (*dynamite* and *machine gun*), and a cornucopia of useful doodads (*stapler, bandsaw, safety razor*) and concepts (*pasteurization*). Like a good department store, you'll find everything you need, whether what you need is a pair of *jeans* or a *lightbulb*. And—also like a good department store—you'll find plenty of things that you probably don't need, like a *cable car* and an *oil drill*.

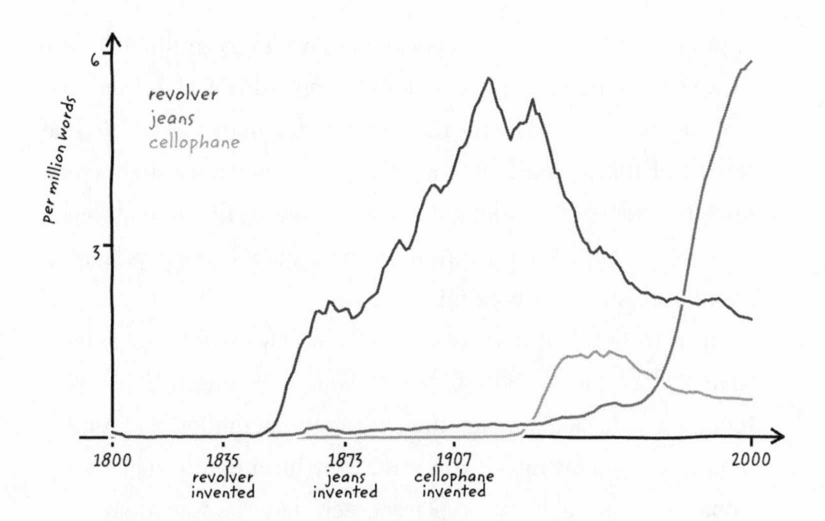

Using this list, we could study the life stories of great inventions. In some cases, like Levi Strauss' *jeans*, the tale is still just beginning: Even today, their impact continues to grow. Other inventions, like *cellophane*, are past their prime. They've taught us something; we might occasionally use them; and their legacy has been passed on to a new generation of ideas. But from the standpoint of our collective memory, they are old hat.

Of course, what's most exciting to us about this list of inventions is that, like the nonsense syllabary of Ebbinghaus, it can give us insight into learning—this time, at the scale of whole societies. In an earlier chapter, we wondered how old the most famous people tend to be when they start making an impact on the cultural record. Now let's ask the same question, but about technology. How long does it take for a given invention to rise to one-quarter of its full cultural impact, as measured by ngrams?

Consider the *revolver*. It was patented in 1835 by Samuel Colt. In 1918, the six-shooter reached peak influence, at a frequency of, appropriately enough, six appearances in every million words. (That's three times as high as *Bill Clinton* at his peak.) It reached 1.5 mentions per million—the one-quarter mark—in 1859. The length of the period between 1835 and 1859, twenty-four years, gives us a sense of how long the *revolver* took to fire up our collective enthusiasm. It's a measure of how quickly society learned about that particular concept.

It turns out that this number varies much more for inventions than it does for celebrities. Sony's *Walkman*, invented in 1978, took only a decade to reach the quarter-impact milestone. Apple's *iPod* was a similar hit—if you want your invention to make a big impact fast, portable music players seem like the way to go. Like the *revolver*, *cellophane* took about a quarter of a century to reach the quarter-impact mark. The *typewriter* took forty-five years. And blue *jeans* took 103. At that rate, Strauss might have made a faster impact as a mathematician.

But these numbers—a century for a new technology to spread—seem very large. Today new technologies routinely transform our daily lives. What's going on? Could collective learning be speeding up?

SINGULARITY OR BUST!

Using ngrams, we can check.

To do so, we combined our Ebbinghaus-inspired list of inventions with Andvord's cohort method. We arranged our 147 technologies by date of invention, starting with the *Jacquard loom* (1801) and ending with the *theremin*, an early electronic instrument (1920). We then grouped them into three periods: inventions of the early nineteenth century, inventions of the mid–nineteenth century, and inventions from around the turn of the century.

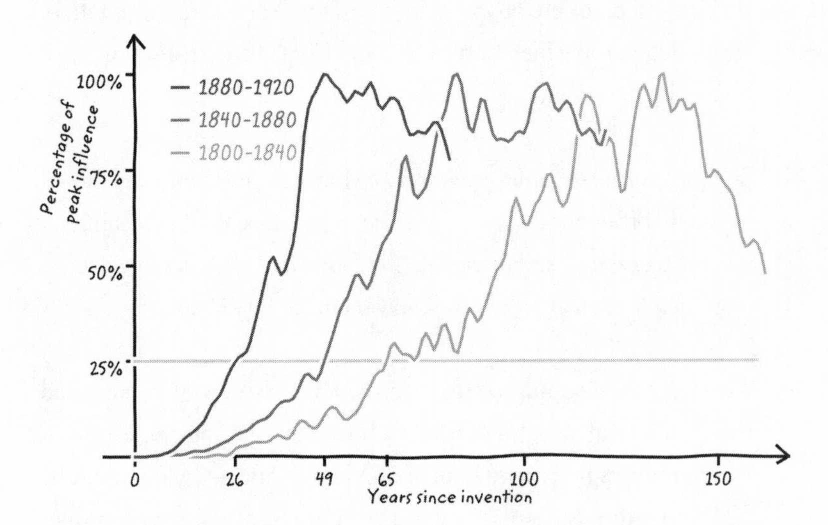

The differences in collective learning over time were obvious. Early-nineteenth-century technologies took sixty-five years to reach the quarter-impact mark. Turn-of-the-century inventions took only twenty-six years. The collective learning curve has been getting shorter and shorter, shrinking by about 2.5 years every decade. Society is learning faster and faster.

Why is that? As with collective forgetting, we don't quite know. But the potential consequences are fascinating to contemplate.

One of the most intriguing possible outcomes of our ever-shrinking collective learning curve emerged from a conversation between the physicist Stanislaw Ulam and the polymath John von Neumann. Ulam was a man who knew about inventions that make a big impact: He invented the hydrogen bomb. Neumann was a famous mathematician, physicist, and game theorist, and a founding father of computer science. (Neumann also coined the phrase Mutually Assured Destruction and its acronym, MAD. Their conversations must have been very fascinating.) Despite his inability to precisely quantify it, Neumann sensed that the rate of technological advancement was increasing. In conversation with Ulam, he observed:

> The ever accelerating progress of technology and changes in the mode of human life . . . gives the appearance of approaching some essential singularity in the history of the race beyond which human affairs, as we know them, could not continue.

This idea was popularized by futurist Ray Kurzweil, who noted that the rate at which computer chips were getting more powerful—a famous regularity known as Moore's law—suggests that, by 2045, an ordinary computer will have more processing power than all of mankind's brains put together. At that point, he predicts that it will be possible to just download our thoughts onto a disk, and to live forever among the machines. This is what Kurzweil refers to as the technological singularity.

This may seem like a strange concept, but Kurzweil is no loony. He sold his first company while a student at MIT and has invented numerous widely used technologies. Bill Gates called Kurzweil "the best person I know at predicting the future of artificial intelligence," and *Forbes* branded him "the ultimate thinking machine." He was awarded the $500,000 Lemelson-MIT Prize in 2001—the world's largest prize for inventors—as well as a National Medal of Technology from Bill Clinton, a man more famous than most of the ingredients in your salad. So there's no doubt that Kurzweil knows his stuff. But is he right?

We really don't know. Ngrams tell us about the past. Alas, they do not predict the future. Yet.

VOLKSGEIST, CULTURE, CULTUROMICS

Our crude measures of memory suggest that it's possible to achieve what the Vienna Circle had thought impossible a century ago: to quantify the spirit of the people, the *Volksgeist*, by empirically measuring aspects of collective consciousness and collective memory.

But what we didn't tell you is that this is a very dangerous endeavor.

Volksgeist is not an innocuous concept. It was introduced, rather innocently, by the German philosopher Johann Gottfried Herder in the eighteenth century. Herder himself was very pluralistic, rejecting slavery, colonialism, and the notion that there were fundamental biological differences between the races. He believed that there were differences between nations—differences that formed what he called *Volksgeist*—but he didn't think that they were a matter of superiority or inferiority.

Yet if you mix the notion of *Volksgeist* with hyperactive nationalism, it's easy to see how Herder's idea can become a fig leaf for racism: I'm superior, because my people have better *Volksgeist*.

In some cases, this is exactly what happened. Think back to what the students claimed in those twelve theses that led to book burnings all over Germany. They "want to respect the traditions of the *Volk*" by eliminating anything that reflected an un-German spirit: *undeutschen Geist*. When it came to matters of racism in the nineteenth and twentieth centuries, the concept of *Volksgeist* was never far to seek.

But there are healthier approaches to *Volksgeist*, too. The German-American intellectual Franz Boas, often called the father of modern anthropology, drew on the very same notion of *Volksgeist* in his work. But he categorically rejected the blending of *Volksgeist* and ultranationalist ideologies, recognizing this dangerous concoction as an intellectually and morally impoverished approach.

Instead, he tried to synthesize *Volksgeist* with the kind of empirical attitude that had motivated Ebbinghaus. To Boas, culture was ever changing but always susceptible to observation and empirical description. By uniting these two traditions, Boas laid the groundwork for the scientific study of culture, creating what we call anthropology today.

It is with Boas in mind that, when speaking to scientists, we like to call what we do "culturomics."

The *-omics* denotes big data, which is what that suffix has come to imply in modern biology and beyond.

The *culture* is the culture of Boas: empirically knowable, its vast variations a matter of endless curiosity and genuine celebration.

CHARTED

2010. In a darkened room at Harvard's Program for Evolutionary Dynamics, a computer chassis stood on a desk, open. Yuan had just come over from Google's Cambridge office, bringing with him hard disks containing the ngram data. The results had finished compiling only hours before. We plugged them in and turned the machine on, eager to confirm that, after three years, we finally had what we thought we had. As the three of us waited for the computer to boot, the only sound in the room was the reassuring whir of the spinning disks.

At last, a command prompt.

Where to start?

Evolution—it's what had gotten us here.

Again, the whir; a minute passed; some more keystrokes followed; and suddenly the command prompt was replaced with a chart. There, through the soft, undulating line, millions of voices spoke to us through the centuries. Drawing from an ocean of data, the curve had distilled a simple, powerful story that anyone could understand.

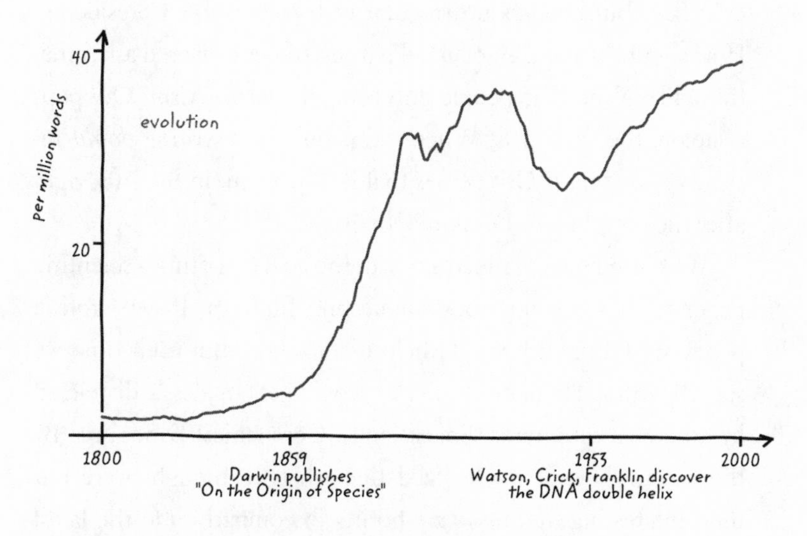

We murmured our approval. Evolution indeed.

The next sound was a pop: the cork.

THE FIRST SAMPLE IS ALWAYS FREE

Once, we had tried to convince someone at Google that building a public tool for studying ngrams, which we proposed calling Bookworm, was a good idea. He quickly took us down a peg or two, responding, "Who's going to use it? Professors. Now, suppose every single professor in the world uses Bookworm. That's, say, one hundred thousand people. At Google, a hundred thousand users doesn't even move the needle."

It was hard for us to argue with that.

But once we got the data and started playing with it, we began to notice something odd: The ngrams were taking over our lives. It was impossible to stop looking at them. We had started with *evolution*. But how about irregular verbs? How about presidents? How about Einstein? At cocktail parties, someone would ask something like: When did people start using the term *sexism*? Out pops a laptop: the early '70s. When did people start writing *donut* instead of *doughnut*? Out comes that laptop again: in the '50s, right after the founding of Dunkin' Donuts.

We started having meetings with the goal of writing a scientific paper to describe our most interesting findings. If we wrote a paper, we thought, it would help us move on. But each time we started writing about one topic, we would get hopelessly distracted by a new set of ngrams. Snack foods! Companies! Dinosaurs! By the meeting's end, we realized that what we thought were our most interesting findings were boring in comparison to the latest eye-opener. It was an impossible situation. How could we manage to break our addiction?

We needed to take a breather, to give ourselves time to gather our thoughts. So we took the four laptops that had access to the ngram database—the only four laptops in the world that could run our prototype Bookworm interface—and started giving them away to other people. One went to Pinker, who quickly began finding charts to include in the book he was writing. Another went to Aviva, Erez's wife. Immediately she reported new discoveries: Checking the German ngram for *Mendelssohn* had led her to start tracking censorship. Now she was addicted, too.

A third machine went to Martin Nowak. When he got home, he casually showed Bookworm to his son Sebastian, who was sixteen years old at the time. Sebastian typed in a query. A chart popped up. Intrigued, he tried another; two queries in, he took the machine away from Martin and excused himself. After ten minutes more, he called a friend: "You have to come over and see this." The friend came over, and the two typed in query after query late into the night.

The last machine went to Google's 2010 Library Summit, where we had been invited to give a keynote address. The summit was where Google typically disclosed the latest news about its digitization project to the heads of many world libraries.

Now, you would think librarians are the quiet type. That was not our experience.

As we explained the basic concept of what we were doing, enthusiasm began to mount—no one had ever heard of anything like this, certainly not at this scale. We had the full attention of every single person in the packed lecture hall. By the time we started showing a few examples, the energy in the room was extraordinary. Finally, after forty-five minutes, we stopped talking and booted up Bookworm. We asked the audience, "Any queries?" We were greeted with thunderous applause, the likes of which we have never heard before or since. But over it, you could hear the librarians begin to shout, unable to contain themselves:

"Try *he* versus *she*!"

"Type in *global warming*!"

"*Pirates* versus *ninjas*!"

The room exploded with excitement, curiosity, glee, and utter fascination.

The ngrams were spellbinding, irresistible, and totally addictive. It was as though we'd discovered a new and extremely nerdy form of heroin.

COPING WITH ADDICTION: A NEW STRATEGY

Sitting in the front row, Dan Clancy could see that the odd little gizmo we had cooked up was going to be as much fun for Google's users as it had been for us and the librarians. He gave the word: Google was going to adapt our prototype and launch it as a part of Google Books. We were thrilled.

Suddenly, our project was transformed from a methodical, scientific tortoise into a Google-powered hare. In two weeks flat, the amazing Google engineers Jon Orwant, Matthew Gray, and William Brockman built a stunning, Web-based version of Bookworm. To avoid the lengthy internal process for approving new trademarks, we had to ditch the name. We gave it a simple, technical label instead: the Ngram Viewer. At 2:00 p.m. on December 16, 2010, the journal *Science* published our research article, and simultaneously, Google launched the Ngram Viewer.

In the first twenty-four hours alone, the site got three million hits. The interwebs were atwitter, and the Twitter was abuzz, with reviews of the Ngram Viewer ranging from "addictive" (@gbilder) to "totally addictive" (@paulfroberts) to "Ohmygoodness the google ngram viewer is the most addictive tool ever" (@rachsyme). *Mother Jones* hailed it as "perhaps the greatest timewaster in the history of the Internet." When we picked up a copy of the *New York Times* the next morning, we were surprised to find our work on the front page.

Problem solved: If we couldn't break our paralyzing addiction to ngrams, at least we could take the rest of the world down with us.

Mommy, where do Martians come from?

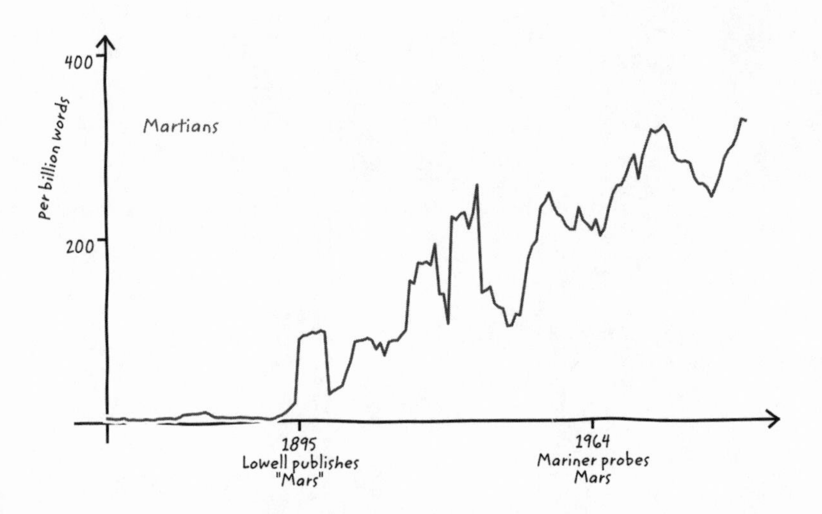

In September 1610, Galileo began a series of observations of the planet Mars. By December of that year, he noticed something remarkable: Mars appeared to be getting smaller and smaller, and was now only a third of its September size. Galileo concluded that, over a period of only a few months, the planet had moved much, much farther from the Earth—a crucial piece of evidence that the Earth was not at the center of the universe. But beyond that, Galileo couldn't see much. His telescope was too primitive to resolve anything about the planet's surface.

Some centuries later, Giovanni Schiaparelli pointed a far more powerful telescope at the red planet. What he saw was remarkable: Massive lines appeared etched into the planet's surface. Schiaparelli's findings so excited a man named Percival Lowell that, in 1894, Lowell decided to build a scope to see for himself. At the observatory he founded in Flagstaff, Arizona, Lowell saw the lines, too. Many members of Lowell's observatory confirmed his findings. On the basis of these direct observations, the team made meticulous maps, showing that the lines formed a dense network crisscrossing the planet.

What could these gargantuan features on the Martian surface be?

Lowell's explanation hinged on the knowledge, already widespread a century ago, that Mars had little water except in the form of ice caps at the planet's poles. Lowell argued that the lines were a vast network of canals, an irrigation system dug by the inhabitants of a dying planet in order to rehydrate their world using water from its polar regions. Based on the lines he saw through the telescope, Lowell concluded that Mars was home to intelligent life. We were not alone.

Among scientists, the arguments about Lowell's work could not have been more heated. Many were skeptical. But some were enthusiastic. Henry Norris Russell, the so-called dean of American astronomers, said of the Martian canals that "perhaps the best of the existing theories, and certainly the most stimulating to the imagination, is that proposed by Mr. Lowell and his fellow workers at his observatory in Arizona."

Lowell's electrifying ideas had an impact far beyond scientific circles. Popularized by a series of three books, they took the world by storm. The breathless news reports came fast and furious. One observer even discovered, embedded in Lowell's canal network, the three-letter

Hebrew name of God: *Shadai*. By 1898, H. G. Wells had written *The War of the Worlds*. Long before the dust settled on Lowell's discoveries, Martians had taken over the Earth—or at least, its imagination.

Scientific enthusiasm for Lowell's ideas had ebbed by the 1910s, in the light of better observations through better telescopes. Still, the half-life of an idea is long, especially such a fun idea, and Lowell's opinions and irrigation maps remained influential. When NASA sent the first unmanned probes to take pictures of the red planet, the Martian globe used to plan the mission was carefully annotated with markings that showed Lowell's canal network. In 1964, as the Mariner probes hurtled through space to their destination, excitement about life on Mars yet again reached a fever pitch.

The pictures that *Mariner 4* sent back on its first flyby of the planet could not have been a greater letdown. There were no canals. No name of God. No obvious signs of intelligent life. Not a single one of Lowell's lines. All that could be seen was a vast expanse of desolate red soil, interrupted by the occasional crater.

The great promise of a new scope is that it can take us to uncharted worlds. But the great danger of a new scope is that, in our enthusiasm, we too quickly pass from what our eyes see to what our mind's eye hopes to see. Even the most powerful data yields to the sovereignty of its interpreter. Martians didn't come from Mars: They came from the mind of a man named Percival Lowell.

Through our scopes, we see ourselves. Every new lens is also a new mirror.

7

UTOPIA, DYSTOPIA, AND DAT(A)TOPIA

In the Book of Samuel, the Israelite king David wonders how many people are under his command. He orders a census. Nine months later, he gets the result: 1.3 million able-bodied fighting men. But David's count angers the Lord, who brings a plague upon the land. For thousands of years, people like David have attempted to quantify aspects of their society. It can be a perilous undertaking.

In this book, we've seen how digital historical records are making it possible to quantify our human collective as never before. Today, we are no longer just counting sheep or counting heads. Instead, we are able to make careful measurements that probe important aspects of our history, language, and culture. And the simple charts we've shown are merely the tip of a vast iceberg. In the coming decades, personal, digital, and historical records are going to totally transform the way we think about ourselves and about the world around us. Before we leave you, we want to sketch out where all this is going and what it's going to mean for science, scholarship, and the quantified society that beckons on the horizon.

And we will grapple, too briefly, with a final question: Is all this a good thing? Will big data turn out to be a promised land? Or could the decisions we make in the coming years come back to plague us?

DIGITAL PAST

The ngram data we've told you about is derived from millions of books. By contemporary standards, that's certainly big data. But when we look back, years from now, we might think differently. After all, a couple million books is just a tiny fraction of our vast cultural output.

Consider a historical figure like Edgar Allan Poe. Unlike many earlier writers, Poe strove to support himself solely via writing. But in the absence of international copyright law, it was hard for a nineteenth-century author to make a living. Driven by pressing financial needs, Poe published his works wherever he could, in an extraordinary array of forums and formats. He wrote poems, short stories, books, plays, novels, reviews, newspaper articles, essays, and letters. He even fabricated a tall tale about a transatlantic balloon voyage, which he managed to get published in a special edition of the New York *Sun*.

When we think of the future of the historical record, and of how digitization will transform it, Poe's works read like a to-do list. Which parts of his oeuvre have made it to the digital commons? How did they get there? And what about the rest? These questions will lead us on a brief, whirlwind tour of the historical record as it exists today.

Books. Our little scope, the Ngram Viewer, was initially powered by 4 percent of all books ever published, or one in twenty-five. In 2012, we helped Yuri Lin, Slav Petrov, and others at Google upgrade the Ngram Viewer to cover about 6 percent of all books, or one in seventeen. Of course, we use only a subset of all the books that Google has digitized. If you include all thirty million of those books, it comes out to a little more than 20 percent of the total. What about the remaining 80 percent? When will they make it into digital archives?

Conveniently, an increasingly large fraction of new books are born digital, distributed as e-books from the moment of publication. Since books are being published in far larger numbers now than at any point in human history, this means that the fraction of books existing in digital form is growing rapidly with each passing day.

That still leaves us with older books, which, somewhat inconveniently, only exist as physical objects. This is where most of the digitization effort will be concentrated. Private corporations and governments are stepping up to the plate, motivated by the desire to both preserve our collective heritage and profit from it. Google continues to lead this effort. It has already digitized more than 30 million of the 130 million books in existence. The company forecasts that it will be done with the rest by 2020. In all likelihood, the vast majority of surviving books will soon be recorded in digital form.

From a quantitative standpoint, this twenty-five-fold improvement in our coverage of the book record, from 4 percent to 100 percent, will make a big difference in terms of the kinds of observations we can make with a cultural telescope. Think again of Galileo, who kicked the Earth out of its perch at the center of the universe with a telescope that was only thirty times better than the naked eye.

Despite this, the study of our book record faces major hurdles.

One serious impediment is copyright law. More aggressive today than it ever was in Poe's time, and just as obsolete, copyright legislation has left the field hamstrung. The Copyright Term Extension Act of 1998 is a good example. This act extended copyright for seventy years after the author's death. It effectively prohibited the online dissemination of nearly all books published after 1923, and it made no provisions for digital research or for the rise of digital libraries. Organizations like the Internet Archive, the HathiTrust, and Project Gutenberg are striving to make books available as openly as possible. But because of the state of copyright legislation, they can do very little about works published in the last century.

This affects the rest of our information ecosystem. For instance, our research group, the Cultural Observatory, has created open-source tools that are far more powerful than the Ngram Viewer, capable of slicing and dicing the book record in all sorts of ways. We can instantly map the usage of the word *raven* across the United States, in works of narrative poetry, written by men in their thirties. But only up to 1923. When it comes to the last century, save if new law affords entry, then the lawyer—dark-robed sentry—who is ever at our door, will yet whisper, "Nevermore!"

And there's another, far more insidious danger that books face. As digital books and digital information become increasingly important, the survival of physical books is being threatened on several fronts. Only three years after introducing the Kindle e-book reader platform, sales of Kindle books at Amazon began to outstrip the sale of printed volumes. And it's not just Amazon: There has been a compelling shift toward e-books in recent years, across a variety of platforms and retailers. In the long run, texts of great importance and sentimental value, like the Bible, will surely remain in print. But such texts are few. For the long tail of this Zipfian distribution, book printing will go the way of the irregular verb. In a few years, books like this one will no longer be printed.

Physical books are also being threatened in what used to be their citadel: the library. The library has, for thousands of years, been the single most important institution working to preserve the historical record. Yet even as online libraries continue to make great strides, their traditional, brick-and-mortar counterparts are facing significant cutbacks. In recent years, 60 percent have faced flat or declining budgets. With funds tight and space even tighter, libraries have no choice but to get rid of old books to make room for new ones. The trouble is that libraries can't just give their old books away. The tracking equipment that is installed in books to keep them from being stolen invariably leads kind souls to find the books and bring them right back. Removing these trackers is too expensive. Instead, libraries are routinely choosing to do something that we might have thought was unimaginable: They are secretly destroying books. This is happening at an astonishing scale. Large libraries sometimes dispose of hundreds of thousands of books at a time.

Which books go? Practices vary from library to library, but it's generally a pretty indiscriminate process. There has been no effort to keep track of what we are losing. In one recent case, volumes from the personal library of the former British prime minister

David Lloyd George were trashed. Occasionally, a library will decide which books to get rid of by just checking which books Google has digitized. The result is an all-out assault on a significant slice of our cultural heritage. A few chapters back we pointed out that censorship can unexpectedly prop up an idea. Here, the opposite is happening: An effort to make books more widely available is threatening the physical survival of those very books. Book digitization will leave a complex legacy.

Newspapers. Of course, the historical record consists of more than just books. Poe's balloon hoax, for instance, appeared in a newspaper. Historical newspapers are an extraordinary resource, reflecting the day-to-day concerns of cities, movements, and other social groups. What are our chances of finding a digital edition of Poe's balloon hoax?

At first glance, we might think the chances are pretty good. Digitization of old newspapers has made significant inroads. Today, major papers like the *New York Times*, the *Boston Globe*, and many others have digitized their full archives. The National Endowment for the Humanities has funded a large effort to digitize Early American newspapers, covering six million pages that span more than a century. Other nations have been making progress, too. Australia's Trove project alone has digitized about one hundred million newspaper articles. Even Google briefly entered the fray, digitizing the archives of two thousand newspapers.

But despite these impressive strides, no newspaper digitization effort is comparable in scale and coverage to what Google is doing for books.

Poe's balloon hoax is a perfect example of this disparity. It is easy to find a digital edition of the hoax today. But that's because of the success of book digitization, not newspaper digitization.

The tall tale is so famous that it appears in many books that anthologize Poe's work. These, along with all of Poe's books, have been digitized.

But you can't find a digital copy of the newspaper that originally printed the story. The NEH has only funded digitization of the New York *Sun* from 1859 to 1920. The hoax, published in 1844, falls into one of the many vast blind spots of newspaper digitization. Most of Poe's newspaper articles have not been digitized, and no one knows when they will be.

Unpublished Text. Publishing itself is a relatively recent invention. Before the printing press, texts circulated as manuscripts, written and copied by hand. Today, a lot of wonderful texts survive only in this form. Many famous manuscripts, like the Dead Sea Scrolls, have been digitized, as have important collections, like the Greek manuscripts at the British Library. But systematic efforts to digitize manuscripts have been fairly local in scope.

Of course, the production of unpublished texts didn't stop with the invention of publication. Poe left 422 letters behind. In his case, the letters have been digitized, but as with his balloon hoax, only because he was so famous that they had been collected in books. Other material by and about Poe has been digitized in Poe-centric efforts, like one at the University of Texas at Austin's Harry Ransom Center. There you can find digital images of some of Poe's original manuscripts, letters that were written to him, and works that he abandoned. You can even see a few Edgar Allan Poe cigarette cards—before baseball cards took over this peculiar cultural niche, cards featuring actors, models, and authors did their part to help sell tobacco.

But when it comes to unpublished material, Poe's legacy isn't very representative. Folks like Poe benefit from a kind of star treatment in the historical record. Anything related to them tends to be tracked down and digitized. What about everyone else? Buried in attics and old trunks, the notes, journals, and correspondence of the 99 percent are usually very hard to get at, and direct efforts to digitize them are the rare exception.

One of the few examples of a successful effort to unearth material of this sort was undertaken by Afsaneh Najmabadi, a Harvard faculty member who studies Iranian women. She went door to door in Iran, asking families if they had preserved any historical documents related to the experience of women. Najmabadi carefully created digital images of everything she found. The result, the Women's Worlds in Qajar Iran archive, is freely available at www.qajarwomen.org. It is a treasure trove of everything from wills to postcards to marriage contracts. All communities have such treasures. But time is slowly leeching them away. Sadly, there is no systematic effort to stop that process.

Physical Objects. Near Poe's old home in Richmond, Virginia, stands the Edgar Allan Poe Museum, where you can see his walking stick, his boyhood bed, some of his old clothes, his wife's piano, a portrait of his foster father, and even a lock of his hair. Such museums remind us that human history is much more than words can tell. History is also found in the maps we drew and the sculptures we crafted. It's in the houses we built, the fields we kept, and the clothes we wore. It's in the food we ate, the music we played, and the gods we believed in. It's in the caves we painted and the fossils of the creatures that came before us.

Inevitably, most of this material will be lost: Our creativity far outstrips our record keeping. But today, more of it can be preserved than ever before. Projects like Europeana strive to make millions of cultural artifacts, drawn from museums, archives, and repositories all over Europe, available in digital form on the Web. Artworks can be photographed at an extraordinarily high resolution, in two or even three dimensions, enabling sites like www.artsy.net to help large numbers of people see some of the world's most important works. Do you really like that piece of Neolithic pottery? Today you can scan it in three dimensions, and use a 3-D printer to print out a replica later.

How much of our history will we capture before it disappears? To make a difference, we need to think big.

We already live in an era of big science. The Large Hadron Collider, and its quest for the Higgs boson, cost $9 billion. The Human Genome Project, whose goal was to determine the sequence of letters that spell out the chemical code underlying human life, cost $3 billion. The amount of money we put into understanding human history is far smaller: The entire annual budget of the National Endowment for the Humanities is about $150 million.

The problem of digitizing the historical record represents an unprecedented opportunity for big-science-style work in the humanities. If we can justify multibillion-dollar projects in the sciences, we should also consider the potential impact of a multibillion-dollar project aimed at recording, preserving, and sharing the most important and fragile tranches of our history to make them widely available for ourselves and our children. By working together, teams of scientists, humanists, and engineers can create shared resources of extraordinary power. These efforts could easily seed the Googles and Facebooks of tomorrow. After all, both these companies started as efforts to digitize aspects of our society. Big humanities is waiting to happen.

Still, despite the vast amount of work left to do, digitization of the historical record has already made significant progress. Having the kinds of resources that we just described available at the click of a button is transforming our appreciation of the past, making it possible to routinely share with our children things that once would have required a trip to the Louvre or the Smithsonian. These resources are going to transform how scientists and humanists approach the past by helping us observe and understand how writing and art, hair and postcards, warfare and romance got to where they are today.

DIGITAL PRESENT

Edgar Allan Poe invented the detective story, a genre whose dramatic engine is the fact that ordinary-seeming people can conceal the darkest of secrets. Suppose you were a historical sleuth who wanted to know Poe's dark secrets: his inner life, his most guarded thoughts. A great place to start would be to look at his personal correspondence. The 422 fascinating letters he left us are just waiting to be explored.

But you know who is a much better-documented writer than Poe? You are. If you are the average American adult, you send 422 e-mails every other week. And you probably have a decade's worth of e-mail living in your account right now. That's hundreds of times as much material as all the correspondence that survives from Poe. And it's not just you who has this fantastic archive: In 2010, two billion e-mailers sent ten trillion e-mails, excluding spam. Today, the average Joe's correspondence is better preserved than the missives of most bygone presidents.

Those e-mail records are a powerful resource. Not only do they document the details of our past, but they also make it possible for us to learn about ourselves in exciting new ways. Take JB's e-mail. A simple ngram analysis of his mailbox can tell you a great deal about JB's life. Over the years, you can see the gradual shift away from French and toward English, reflecting acculturation to the United States after he moved from France. Friendships come and go. Youthful enthusiasms fade: *party* decreases in frequency over a decade. At the same time, his love life unfolds, converging on a final ngram: *Ina*. Exploring his ngrams in this way, JB repeatedly rediscovered things that had once been important to him, but were slowly forgotten. Big data doesn't have to be daunting. It can be an intimate window into our own lives. Into our quantified selves.

Our digital memories extend far beyond correspondence. Along with fifteen thousand e-mails, the average person sends or receives five thousand e-mail attachments each year. They "like" about 140 things. They upload eighteen pictures to Facebook, and two more to Instagram. They tweet nine times. They put up twenty seconds of video on YouTube. They upload fifty-two files to Dropbox. They interact with forty-three friends on an online social network. And these impressive averages don't account for all the images, documents, videos, and music that we create but don't share online. And they don't account for the fact that nearly three-quarters of the world's population still lacks Internet access.

Taken together, this material comprises an astonishingly detailed record of the lives of billions of people—a record that did not exist at all mere decades ago. It has no precedent in human history. Our civilization tweets more words every hour than can be found in all the surviving texts of ancient Greece. Compared to the average person today, a man like Poe is an enigma.

Yet compared to the people of tomorrow, the people of today are a total mystery.

DIGITAL FUTURE

At the beginning of this book, we told you that the average person alive today produces a little less than one terabyte of data each year. But some people are above average. One of these people is Dwayne Roy, a toddler living in Boston. He regularly produces that much data in a single weekend.

Why does Dwayne produce so many bits? Dwayne is the son of Professor Deb Roy, who runs the Cognitive Machines Group at the MIT Media Lab, and Professor Rupal Patel, who studies speech pathology at Northeastern. Both are fascinated by how children learn to speak. She cares, because it's exactly what her discipline is about. He cares, because he wants to use the same principles to teach robots how to communicate in ordinary human language. The couple realized that one of the central challenges in understanding how children acquire language is a lack of data. No one had documented, in detail, all the ways in which children are exposed to language as they grow up.

When Patel became pregnant, the pair decided to tackle this problem head-on by comprehensively recording the first three years of their new baby's life. Funded by a grant from the National Science Foundation for what Roy called the Human Speechome Project, he outfitted their family's home with eleven high-resolution video cameras and fourteen microphones. Three thousand feet of cable connect these devices to a data center that lives in their basement. Each day, this basement outpost stores more than three hundred gigabytes of information about Dwayne. Every step he takes, every noise he makes, every sound he hears, and every sight he sees—all of it is recorded for the benefit of science. (The cameras shut down when the baby is asleep, and obviously can't track him when he's out of the house.)

With so much information pouring in, the basement data center tends to flood. That's why the elder Roy has to regularly take suitcases full of hard disks to be permanently archived on a far more powerful computer system that he has constructed at work. To track one small boy, he uses a multimillion-dollar CPU grid outfitted with a massive disk array capable of storing a petabyte, or one million gigabytes. The system's name doubles as its job description: TotalRecall.

Today, Dwayne Roy is an exception. Not everyone is the subject of an attempt to record and preserve a video feed of his entire life. But as digital media and human life interpenetrate ever more deeply, this sort of record will become commonplace.

We can already see the types of devices that will usher in this transformation. Google recently introduced Glass, an eyeglass-mounted reality augmentation system that features a webcam tracking everything in your field of view and a small monitor to provide you with relevant information about what you're seeing and doing in real time. Baking a cake? The glasses might figure that out, find the recipe, and show you instructions as you progress. Don't recognize that guy who just walked up to you? No problem—using face recognition, Google Glass could remind you. Sure, the glasses look a bit silly. But do you remember how silly people looked, talking aloud to themselves, in the early days of the cell phone? Whether or not Google Glass ever takes off, this kind of technology is certain to have a bright future.

Such devices make Dwayne Roy–grade life logging easy. At first, almost no one will be interested in doing that sort of thing—it is the ultimate breach of privacy. But the Internet has been redefining privacy norms from the beginning, inducing people to broadcast an ever-increasing amount of personal information, whether it be blogging our daily thoughts or announcing our relationship status. We know how this story ends: Inevitably, some people will voluntarily start to record their entire lives, and Web sites will pop up to help them with the distribution.

There are some obvious benefits. With a life log, you'd never irretrievably forget anything—you could just look up every sensory experience you've ever had. That can be a good thing. (Sometimes.) It might make you safer, too. After all, who would harm someone if the crime was being aired live? You could have real-time life coaching, with people around the world giving you nonstop advice about what to do next. (On second thought, that could rapidly become annoying.) Occasionally, you might go offlog, disabling your life logger for an intimate moment or a bathroom break. Most people will probably do this. Some will not.

Life logging will be as much a window on our bodies as it is a window on the world we inhabit. Wearable electronics like the Nike+ FuelBand and the Fitbit already keep track of how many steps you've taken, how many stairs you've climbed, and how many calories you've burned, all day long. A gadget called the Scanadu Scout is more ambitious: A small, handheld disk, the Scout tracks and records your body temperature, heart rate, and blood oxygen levels, all in seconds. It can also perform an electrocardiogram and even analyze your urine. Basically, the Scout is humanity's first draft of a *Star Trek*–style tricorder. Such data will ensure that life logs also serve as medical records, saturated with details about all the unconscious processes that keep our bodies going. If something goes wrong, the log will immediately notify caregivers. Today's paradigm of visiting the doctor for an annual checkup will be turned on its head. Using tricorder-based telemedicine, health care providers will be able to track how you're doing all day, every day. If something seems amiss, they'll be just as likely to call you as you are to call them.

Life logging will allow us to record a staggering fraction of what happens to us, both inside and outside our bodies. But what about the most evanescent of all experiences: human thought?

We think that the mind-reading gizmos of science fiction, capable of involuntarily transcribing a user's every thought, are unlikely to become a reality anytime soon. The problem is that it is hard to train a machine to make sense of ordinary brain waves. But there may be a powerful work-around. In the last decade or so, scientists have been successfully developing mind-machine interfaces that enable paralyzed individuals to move a prosthetic limb with the power of thought, or to wirelessly broadcast a mental command that moves a computer mouse. Such interfaces have been used to communicate with people who appear, by the ordinary medical definition, to be comatose. They're even making their way into toys.

These interfaces rely on the fact that, although ordinary brain waves are confusing to a mechanical eavesdropper, we can train our brains to make their activity more transparent to a machine. This is accomplished by voluntarily generating specific neural cues that a machine is capable of recognizing. In every such interface—whether it's an fMRI scanner tracking blood flow in the brain, an electroencephalogram tracking electrical activity, or a neural implant linked to a small cluster of brain cells—all the machine does is look for an agreed-upon signal and respond to it in a preprogrammed way. This approach has been enormously successful. It's not hard to imagine such systems allowing all of us to use our minds to operate appliances, or even send messages to one another. And that may be just the start.

When we think, our cogitation frequently takes the form of a sequence of words. There's a special phrase we use to describe this phenomenon: the stream of consciousness. On some level, the existence of the stream of consciousness is surprising. Words are a system for communicating with other people. It's not obvious why we also use them to organize our internal thoughts when no other person is involved. But we all do.

From the brain's point of view, a neural cue to a mind-machine interface is not so different from a spoken word. It's all just brain cells firing in patterns. The main difference is that, instead of using this neural word to talk to a person, we use it to talk to a machine. It's not crazy to think that people might get used to accompanying their internal monologue with the corresponding mindwords, creating a real-time closed-captioning system for the benefit of the machines in their audience. By cooperating with computers in this way, it might be possible to log our inner monologue.

Every sensory experience, every beat of our heart, every rumble in our stomach, and even every thought that crosses our mind—all these are in principle loggable. Actually logging them will change our lives in breathtaking ways that we can hardly imagine today. And these logs won't just change our own lives. If we so choose, our life logs will outlive us. We will be able to leave a complete chronicle of our existence to children and loved ones. They will be able to learn from our triumphs and our regrets, our wisdom and our foolishness: a digital afterlife. If you were so inclined, you could sell your life log to a company, or share it with scientists and scholars. In the library of the future, the biography section won't just have the stories of people's lives. It will have the complete broadcast.

TRUTH AND CONSEQUENCES

On April 15, 2013, two bombs exploded two hundred yards from the end of the Boston Marathon. Shrapnel tore through the massive crowds that had assembled at the finish line. Three spectators were killed. Hundreds were wounded. At least fourteen victims required amputations. In the days following the event, the FBI was desperate for clues, but there was little evidence. The bombs had been constructed from pressure cookers, hidden in backpacks, loaded with nails, ball bearings, and scraps of metal. All of these items can easily be obtained by anyone. Half a million spectators had watched the race. Which of them had planted the bombs? It was a whodunit at the largest scale imaginable.

But the FBI had a powerful trick up its sleeve: digital history. The Bureau recognized that, in one respect, the massive number of people at the scene of the crime was an advantage. Spectators take photographs. The stores that lined the street had their own cameras, too. With so many cameras in such a small space, and so many pictures being snapped in such a short period of time, surely someone would have taken a good photo of the culprit holding the backpack.

That hunch was right, and within days, the investigators released images from a Lord & Taylor surveillance video in which the bombers—two, it turned out—could clearly be seen. Tips started streaming in, many in the form of high-resolution photos that had, by sheer coincidence, captured the suspects' faces. With their pictures spreading quickly across the Web, the bombers went on a final, bloody rampage. One was killed in a shoot-out with police. The other was caught. But their plans for additional bombings—they had intended to attack New York's Times Square next—came to nothing. Bad guys be warned: Whoever you are, wherever you are, big data can track you down.

But digitized history does more than hound the bad guys. It can also hurt the innocent.

In November 2011, fifteen-year-old Rehtaeh Parsons went to a party, where she was allegedly raped by four boys. The boys took pictures. The pictures began to spread over e-mail and on Facebook. Instead of rallying around her, Parsons' peers made her life into a nightmare. Faced with constant bullying, she changed schools. Her family moved. She was hospitalized for weeks at a time. But there was no escaping the shame. There was no escaping the bullying, both online and off. There was no escaping those digital pictures that would never go away. In April 2013, Parsons hanged herself.

DATA IS POWER

Photography, from its inception, has been dogged by a somewhat peculiar superstition: that by recording your image, a camera steals a tiny part of your soul. There is something to that idea. As we just saw, having just a single picture of someone can give you a form of power over that person. Will big data steal your soul outright?

This is an urgent question. Because it used to take a deliberate effort to preserve something for posterity, very little was recorded. But we've come a long way from carving our data on a rock. Soon it will become so easy to track much of what we experience that many of us will find it simpler to just record everything by default. It will take a deliberate choice to keep something off the record. As a result, preserving information is changing from a technological puzzle into a moral dilemma. And the dilemma turns on a small handful of issues. What are the things that belong offlog? And if there is a log, who has the right to access it?

It's hard to tell how these questions will be answered, because it is much easier to speculate about the future of our technology than about the future of our values. Take Dwayne Roy's case. Even if the motivation is to advance science, is it really right that a two-year-old boy has less privacy than the president of the United States? Many people would object to being documented in that way. But the social web is transforming communal norms about sharing at a shocking pace. Lots of the things that we share online today would have been closely guarded twenty years ago, or even five years ago. Perhaps the kids of Dwayne's generation won't mind. Perhaps they will all think it hopelessly primitive not to have a life log of one's formative years.

Still, call us old-fashioned, but just as it seems apparent that life logging will become possible, it seems equally apparent to us that public life logs are a very dangerous concept. Marketers, of course, will use them to continue flooding us with annoying advertisements. Already, the chain store Target can use its data analytics to figure out which of its customers is pregnant. On one occasion, Target coupons broke the news of a teenager's pregnancy to her unsuspecting parents. One can only imagine how unpleasant this would get if marketers and global corporations had unregulated access to life logs.

Yet corporate interference may not be the worst of our concerns. A government could use life logging to track all citizens, all the time. Already, companies like Google and Facebook open their records to the federal government when national security is at stake. Sometimes, the government manages to get at the records whether the company likes it or not. In September 2012, Twitter was forced by a New York criminal court to hand over the private tweets of Malcolm Harris, one of the Occupy Wall Street protesters. In 2013, the Edward Snowden leaks unleashed national outrage, prompting President Obama to reassure Americans that "nobody is listening to your telephone calls." Where is the line between legitimate public interest and Big Brother? It must exist. In a world where the government can subpoena anyone's life log, anytime, resistance really is futile.

Worse still are the dystopias one can imagine if mind logging ever becomes technically feasible. Here's one: A totalitarian government might force everyone to log every thought, all the time. Blank entries in the mind log would be punished, and private thoughts would become a thing of the past. That's not even the most terrifying scenario. Imagine if a government enforced a mandatory mind log, requiring citizens to transcribe specific thoughts, over and over, the way schoolchildren might recite the Pledge of Allegiance or a catechism. Trapped in a compulsory stream of consciousness, citizens would become prisoners of their own minds.

These are huge concerns. But even though life logging is still only a nascent possibility, one can already begin to see the seeds of a countermovement. In Seattle, the owners of the 5 Point Café worry that the presence of life-logging technology will discourage customers from engaging in their typical, freewheeling shenanigans. An absence of shenanigans would obviously be bad for business, so the bar has banned Google Glass. A Web start-up called Snapchat offers a service allowing users to send messages that are deleted after a specified length of time. As life logging becomes increasingly common, it will create the need for offlog spaces, offlog times, and offlog interactions.

Our lives cast digital shadows. The battle for those big shadows, the right to own our personal history and to control who has access to it, is already met. Will the digital commons grow up to be a vast and wondrous playground? A powerful tool for law enforcement? The experiential and moral legacy of countless generations? Or the backbone of a surveillance state? This contest will be one of the great moral conflicts of the coming century.

KINDRED SPIRITS

Galileo's telescope—two lenses, back to back—marked a turning point in the history of our civilization. What he saw contradicted Catholic Church doctrine. For his trouble, the Inquisition put him under house arrest, where he remained for the rest of his life. But the Church could not arrest his ideas. After Galileo—and in no small part because of him—the Church's lengthy dominion over the Western mind began to ebb.

In its place, two great intellectual traditions took root. One was the sciences, tasked with determining the nature of the universe by means of empirical observation. The other was the humanities, the study of human nature through careful, critical analysis. Together, these two siblings have given many powerful gifts to Western civilization, from freedom and democracy to engineering and technology.

Yet these mighty brethren have long been estranged. Even today a typical student must choose to focus on either the sciences or the humanities; rare is the major or degree program that spans the two. A typical researcher, too, must ally with one group or the other. The boundaries have long been encoded into our schools, our universities, and our entire knowledge ecosystem. We study math. We study Shakespeare. But not together.

At least not until recently. At Stanford, an Italian scholar named Franco Moretti has started using the onslaught of digital books to study the interaction network of characters in Shakespeare, applying methods and approaches from computer science and statistical physics in a radically new domain. Matthew Jockers, a literature professor at the University of Nebraska, is able to identify subtle relationships between nineteenth-century novels based on things as seemingly esoteric as the statistical distribution of the pronouns that they contain. At the National Endowment for the Humanities, Brett Bobley heads an innovative program called the Digging into Data Challenge, which helps humanists all over the United States think critically about what all this new data can do for them. They are going where no math has gone before.

Except at Dartmouth, where a mathematician named Daniel Rockmore has been using digital books to study how authors' styles influence one another. He uses much more math than Moretti, and much less reading. But the two are kindred spirits. Or at the University of Texas at Austin, where psychologist James Pennebaker has been studying how the distribution of pronouns in a text reflects the mood of the author. Pennebaker and Jockers come from completely different intellectual traditions, but they, too, are kindred spirits. Or at the White House's Office of Science and Technology Policy, where Tom Kalil is spearheading a big data initiative at the behest of President Obama himself. Kalil and Bobley don't fund the same people. But they are kindred spirits as well.

As the nature of the historical record changes, it is scrambling the boundaries between science and the humanities. The resulting mishmash goes by many names. Historians who do this sort of thing are apt to call themselves "digital humanists." Linguistics departments have "corpus linguists." Psychologists and sociologists sometimes prefer the term "computational social scientist." And in one Silicon Valley start-up after another, this simmering conceptual chulent is just business as usual.

Slowly, minds from all sides of this deep rift are coming together. At an academic conference in Maryland during the spring of 2013, the National Institutes of Health, the National Endowment for the Humanities, and the National Library of Medicine convened a group of researchers spanning an astonishing range of disciplines, from art history to African languages to computer science, from microbiology to rhetoric to poetics to zoology. David Searls, former senior vice president at pharmaceutical giant Glaxo-

SmithKline, gave the keynote address. It was the first time that the NIH and the NEH had ever gotten together to sponsor a conference. The topic, "Data, Biomedicine, and the Digital Humanities," betrays an astonishing optimism: the idea that historians and philosophers and artists and doctors and biologists, thinking about data together, can advance their individual causes better than any of them can alone. The conference title, "Shared Horizons," was dead-on. At the interface of all our work lies the most exciting terrain in our intellectual future.

No one knows quite what to call it. And no one knows quite where it's going. But one thing is certain: Science and the humanities are becoming, once again, kindred spirits. And just as Galileo transformed our understanding of our world in the seventeenth century, these two lenses, back to back, will do the same in the twenty-first.

PSYCHOHISTORY

Gaal Dornick, using nonmathematical concepts, has defined psychohistory to be that branch of mathematics which deals with the reactions of human conglomerates to fixed social and economic stimuli. . . .

Implicit in all these definitions is the assumption that the human conglomerate being dealt with is sufficiently large for valid statistical treatment. . . . A further necessary assumption is that the human conglomerate be itself unaware of psychohistoric analysis in order that its reactions be truly random. . . .

—Isaac Asimov, *Foundation*

In one of the most famous books in all of science fiction, *Foundation*, Isaac Asimov imagines a mathematician named Hari Seldon. Seldon's great contribution is that he figures out how to predict the future by combining elaborate mathematical theories with detailed measurements about the state of society at any given moment in time. Of course, Seldon can't know what a particular person will do: Individual people are too random. But he can figure out what society as a whole will do. For instance, Seldon figures out that the Empire, which has ruled the galaxy for more than a millennium, will soon fall. Seldon's theory doesn't tell him exactly who will do exactly what to bring about the fall, but it does tell him that the fall is imminent, and that it will leave chaos in its wake.

Such theories of aggregate behavior are not uncommon in the sciences. Consider what happens when you inflate a balloon and then, without tying the knot, let go. A small child learns that air will start flowing out of the opening and that, as the balloon deflates, it will fly away, eventually falling to the ground. A physicist could do better, calculating the rate at which air molecules spill out of the hole, the pace of deflation, and the speed of the balloon as it whizzes through the air. But no scientist in the world can tell you in what order the individual gas molecules in the balloon will hurtle out: Single molecules are far too random. The balloon, along with the air it contains, follows a predictable pattern, but only when considered in aggregate.

Asimov's idea—which he dubbed psychohistory—was that such an approach might make it possible to predict the future, in aggregate, of human civilization.

To a contemporary social scientist, this enthusiastic brand of cultural determinism may seem utterly foreign. It's a notion that most fields—economics is a notable exception—give little credence. That's a bit surprising, because Asimov's concept is actually the ur-doctrine of social science. In the early nineteenth century, Auguste Comte, the father of sociology and the founder of the social sciences, believed that careful empirical study would eventually reveal the laws that governed the operation of human society, in the same way that careful study of physical phenomena had revealed underlying mathematical principles. His original name for the discipline that he later dubbed sociology was social physics. Comte believed that understanding the laws of sociology would make it possible to use them to create a better society, much in the way that an understanding of physics can be used to build a better toaster. When Asimov's Hari Seldon, on the basis of psychohistorical calculations, takes actions to minimize galactic chaos, he is the fictional embodiment of Comte's fantasy.

It is very tempting, when thinking about the tidal wave of data that will soon break over the social sciences, to imagine that, with so much data, Comte's dream might be within reach.

On the other hand, attempting to predict historical trends before they happen seems completely nuts.

So, using ngrams, we decided to do one last experiment, whose goal was to check whether historical trends might be predictable. We tested the simplest possible prediction, something we call cultural inertia. All we mean by cultural inertia is that ngrams that are going up will tend to keep going up, and ngrams that are going down will tend to keep going down. The stock market doesn't exhibit inertia: If it did, anyone could make a killing as an investor. If human culture exhibits inertia, then we can learn a lot about what an ngram will do next by examining what it just did.

Here is the chart that the robot drew:

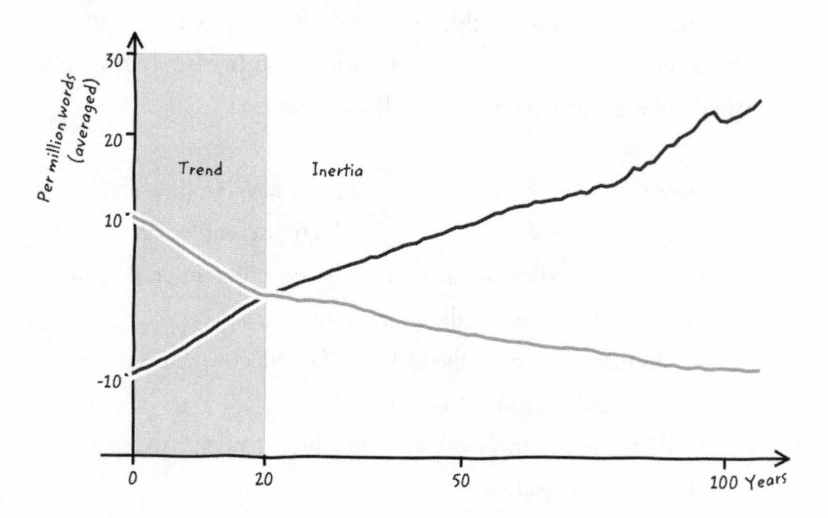

In light gray, we show the average frequency of a large number of ngrams that were chosen for inclusion because they show a consistent decline over a twenty-year period. Does the trend continue when the period has ended? It does, for decades thereafter. In dark gray, we examine the reverse, a collection of ngrams that increased consistently over a twenty-year period. Their dramatic ascent continues for nearly a century—for as long as we can measure. So there you have it: Ngrams that are going up tend to keep going up. Ngrams that are going down tend to keep going down. More generally: Ngrams in motion tend to stay in motion (unless acted on by a psychohistorical force).

Maybe, just maybe, a predictive science of history is possible.

Maybe, just maybe, our culture obeys deterministic laws. And maybe, just maybe, that is where all of our data is taking us.

But even if such an understanding is possible, is it really what we want? Comte thought so. He believed that without objective measurement, without falsifiable predictions, our understanding of human history, society, and culture would be deeply impoverished. The anthropologist Franz Boas disagreed:

> The physicist compares a series of similar facts, from which he isolates the general phenomenon which is common to all of them. Henceforth the single facts become less important to him, as he lays stress on the general law alone.
>
> On the other hand, the facts are the object which is of importance and interest to the historian. . . .
>
> Which of the two methods is of a higher value? An answer can only be subjective. . . .

In short: Sometimes, you want to look at a chart. Other times, you want to curl up with a good book. Welcome to history in our digital future. Why not try both?

APPENDIX

GREAT BATTLES OF HISTORY

Dilemmas

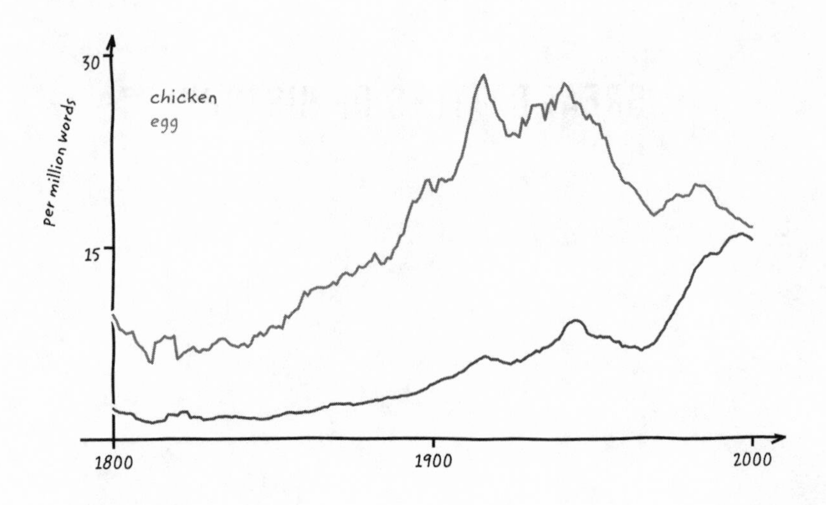
30
15
per million words
chicken
egg
1800
1900
2000

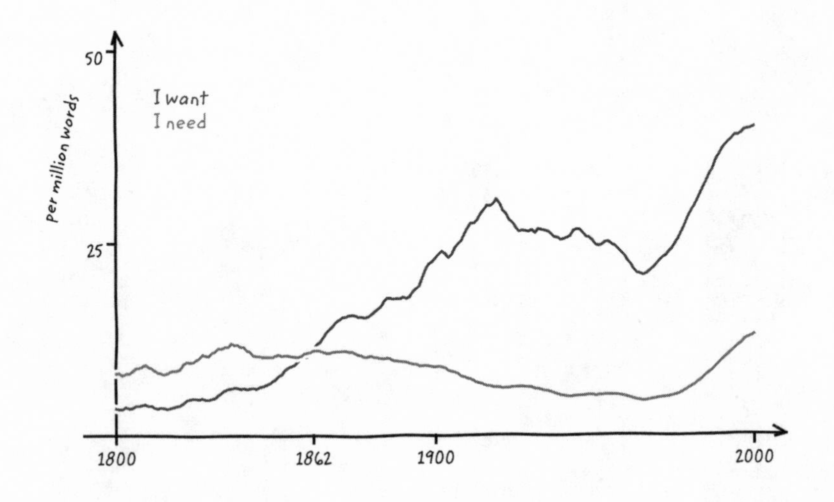
50
25
per million words
I want
I need
1800
1862
1900
2000

Duels

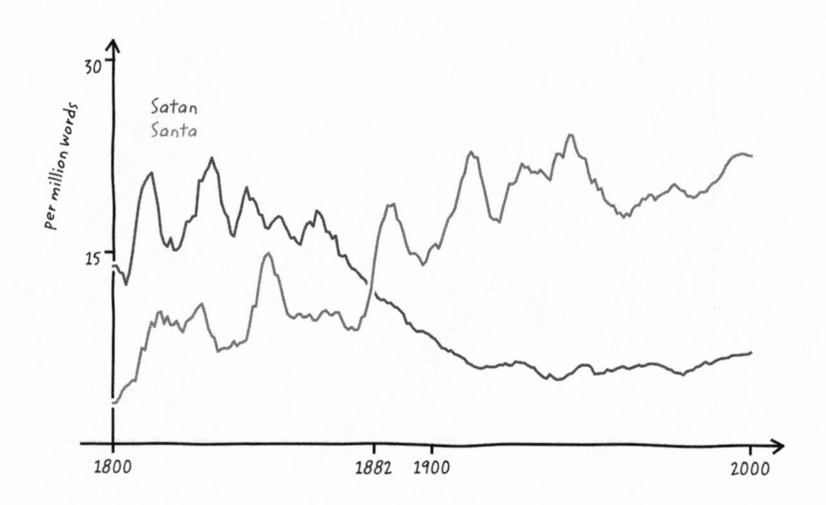
30
15
Per million words
Satan
Santa
1800
1882
1900
2000

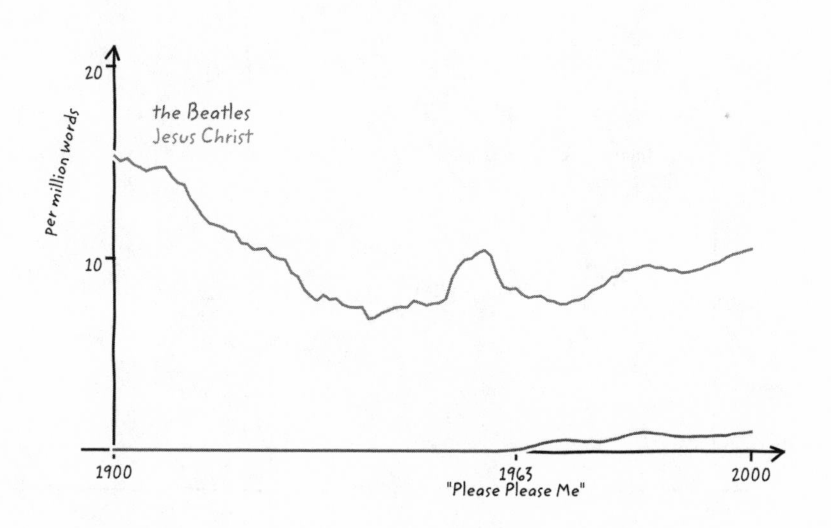
20
10
Per million words
the Beatles
Jesus Christ
1900
1963
"Please Please Me"
2000

Religion

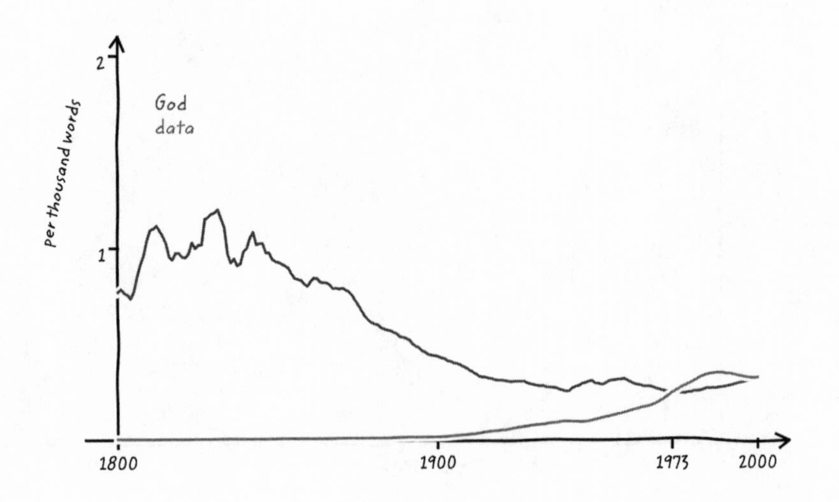
God
data
per thousand words
2
1
1800
1900
1973
2000

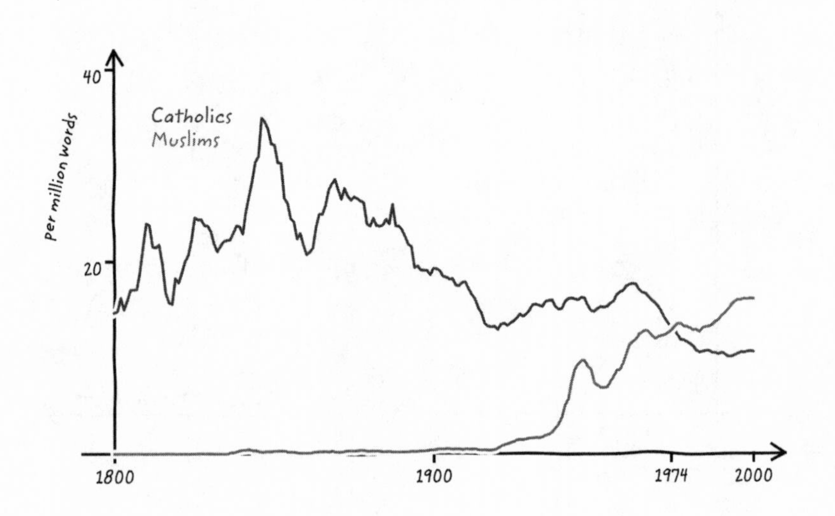
Catholics
Muslims
per million words
40
20
1800
1900
1974
2000

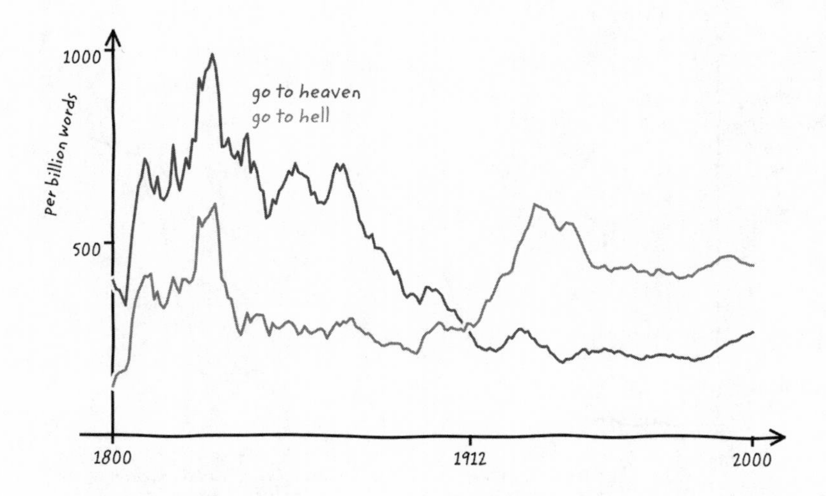
1000
500
per billion words
go to heaven
go to hell
1800
1912
2000

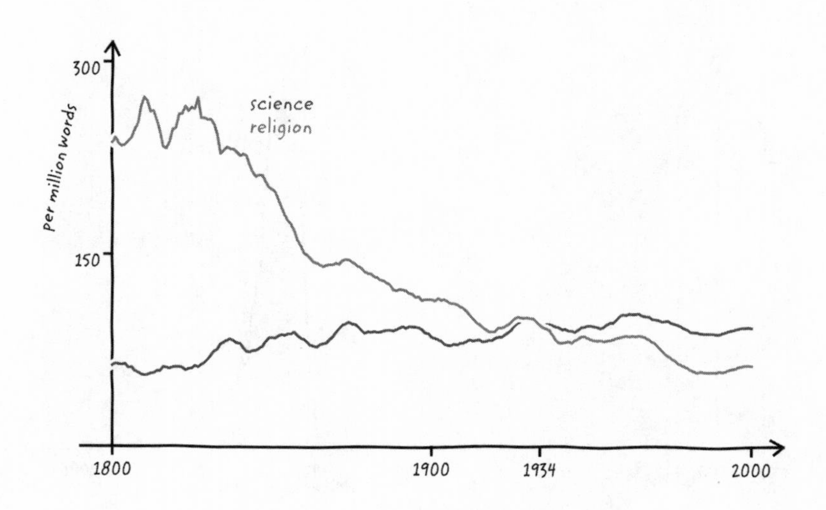
300
150
per million words
science
religion
1800
1900
1934
2000

Science

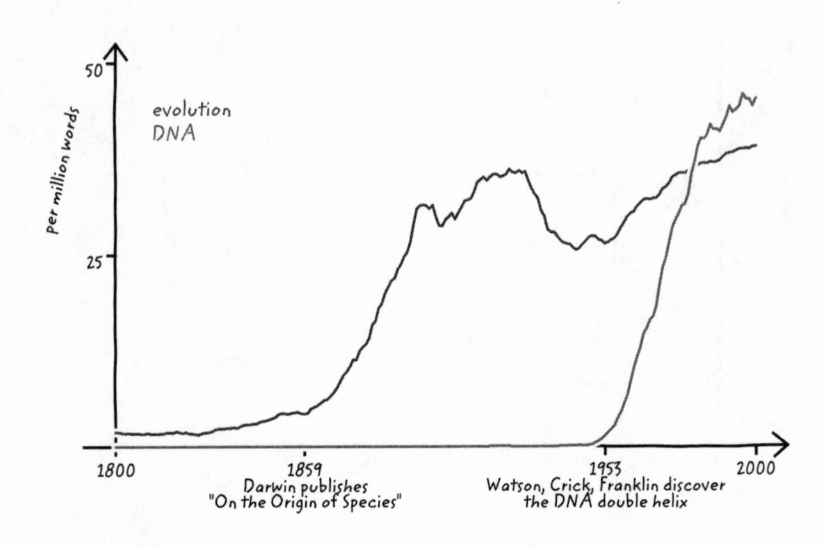
50
25
per million words
evolution
DNA
1800
1859
Darwin publishes
"On the Origin of Species"
1953
Watson, Crick, Franklin discover
the DNA double helix
2000

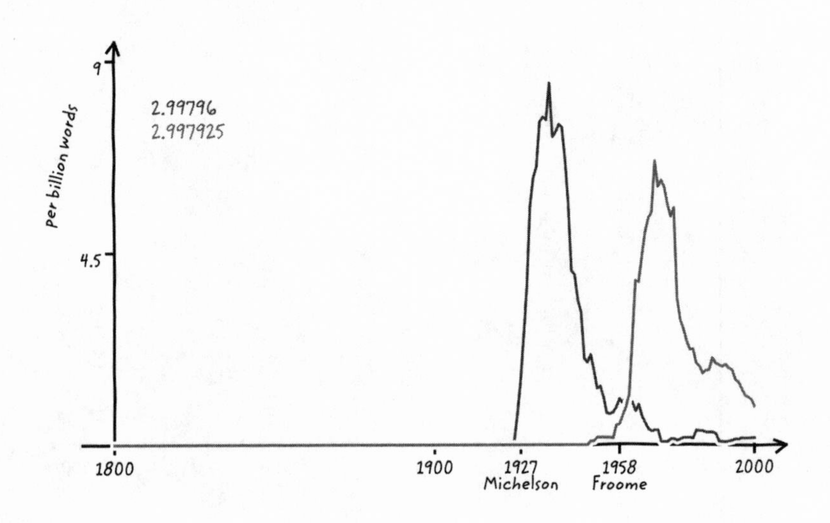
9
4.5
per billion words
2.99796
2.997925
1800
1900
1927
Michelson
1958
Froome
2000

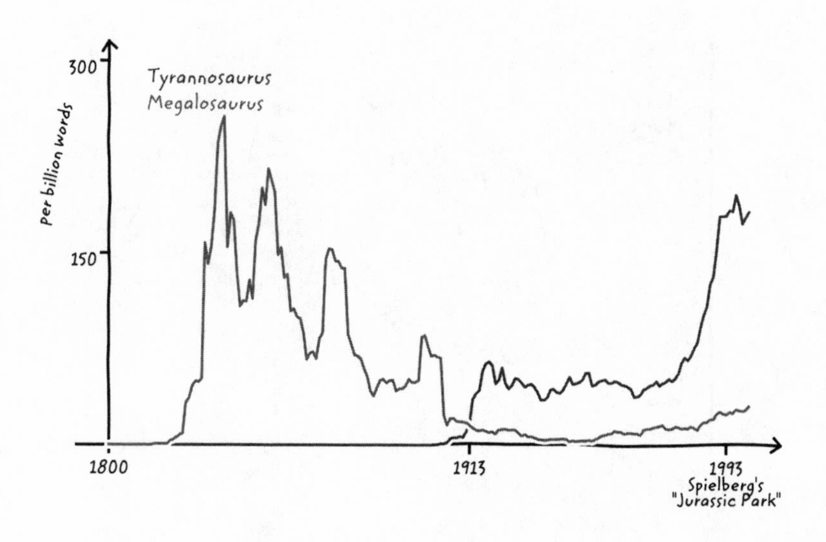
300
150
per billion words
Tyrannosaurus
Megalosaurus
1800
1913
1993
Spielberg's
"Jurassic Park"

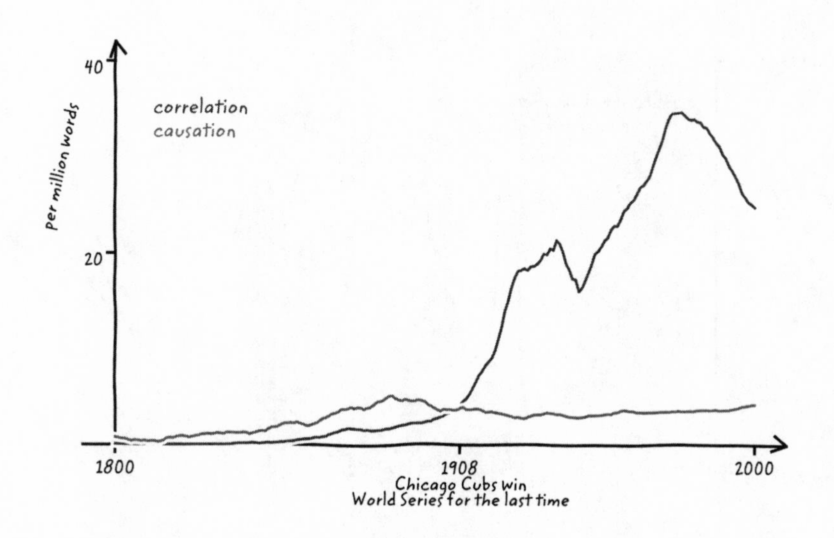
40
20
per million words
correlation
causation
1800
1908
Chicago Cubs win
World Series for the last time
2000

Politics

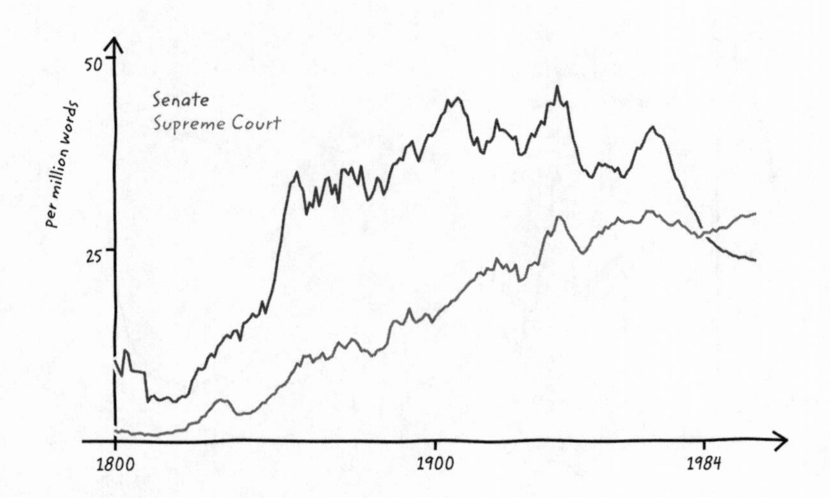
50
25
per million words
Senate
Supreme Court
1800
1900
1984

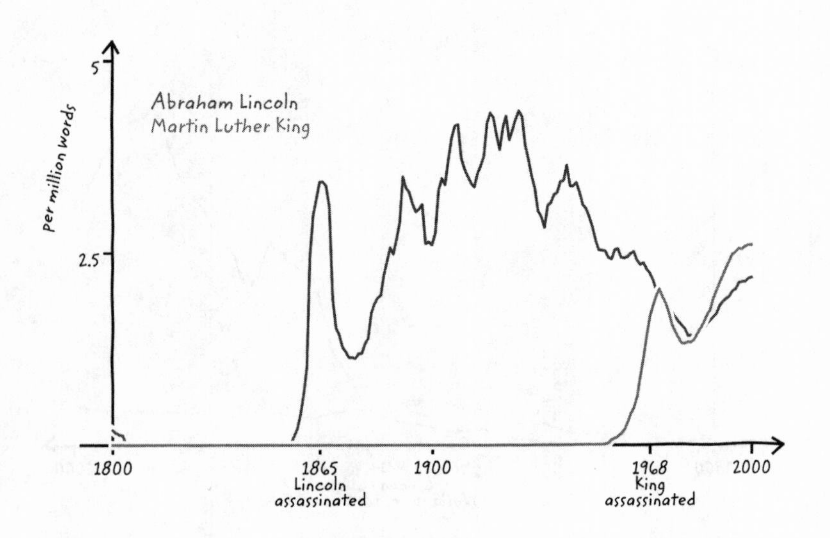
5
2.5
per million words
Abraham Lincoln
Martin Luther King
1800
1865
Lincoln
assassinated
1900
1968
King
assassinated
2000

Social Change

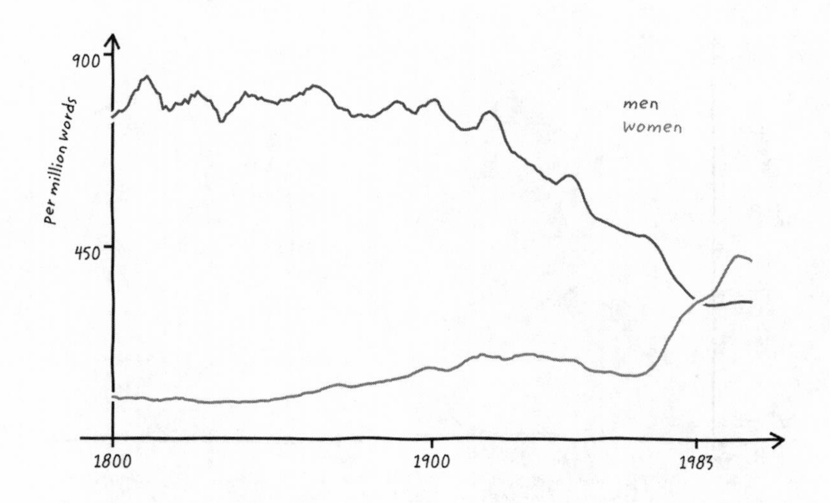

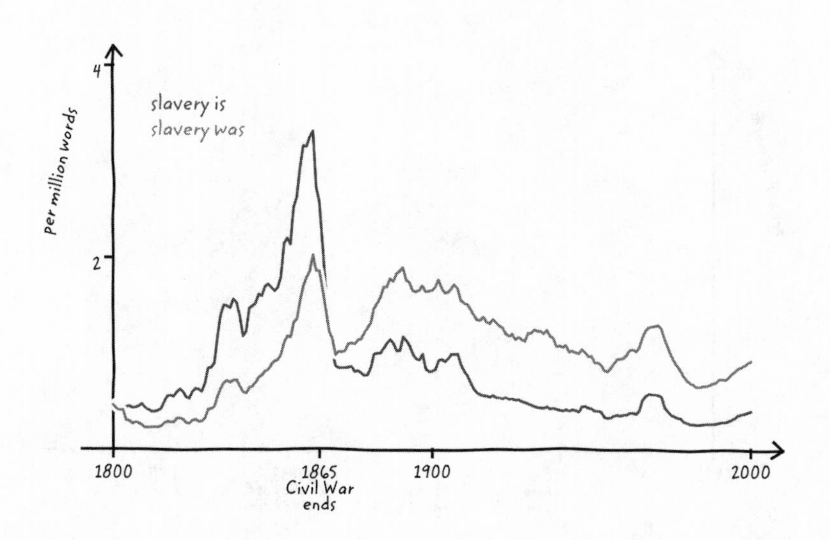

Economics

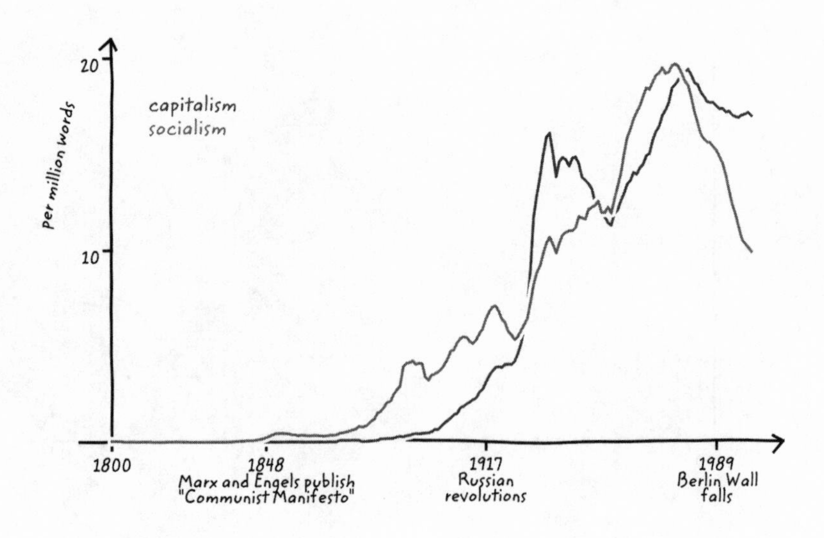

20
10
per million words
capitalism
socialism
1800
1848
Marx and Engels publish "Communist Manifesto"
1917
Russian revolutions
1989
Berlin Wall falls

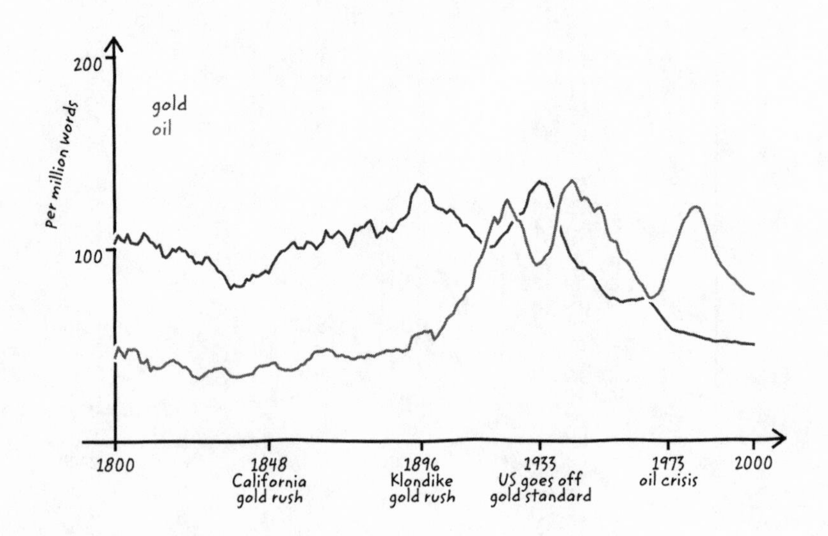

200
100
per million words
gold
oil
1800
1848
California gold rush
1896
Klondike gold rush
1933
US goes off gold standard
1973
oil crisis
2000

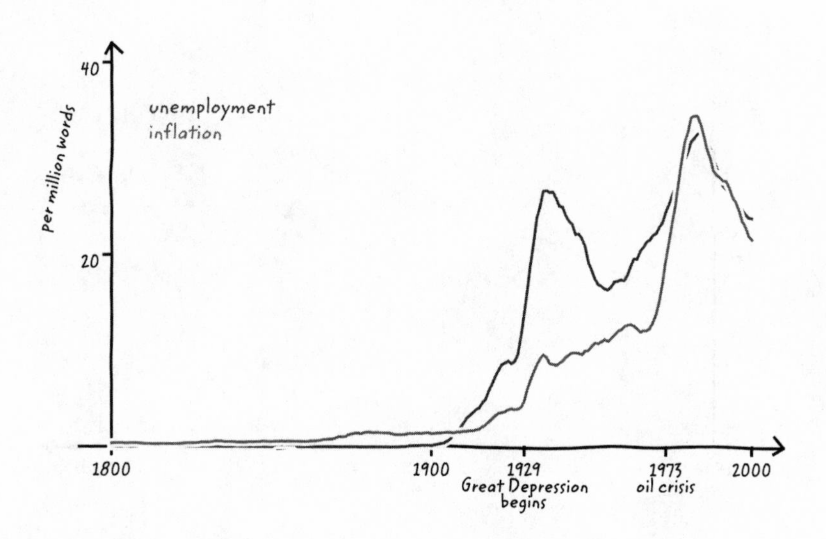
per million words
40
20
unemployment
inflation
1800
1900
1929
Great Depression
begins
1973
oil crisis
2000

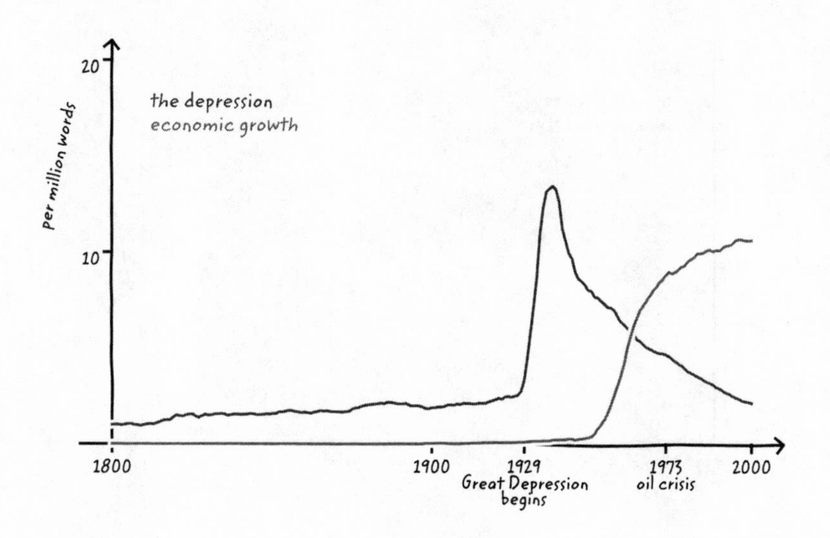
per million words
20
10
the depression
economic growth
1800
1900
1929
Great Depression
begins
1973
oil crisis
2000

The Environment

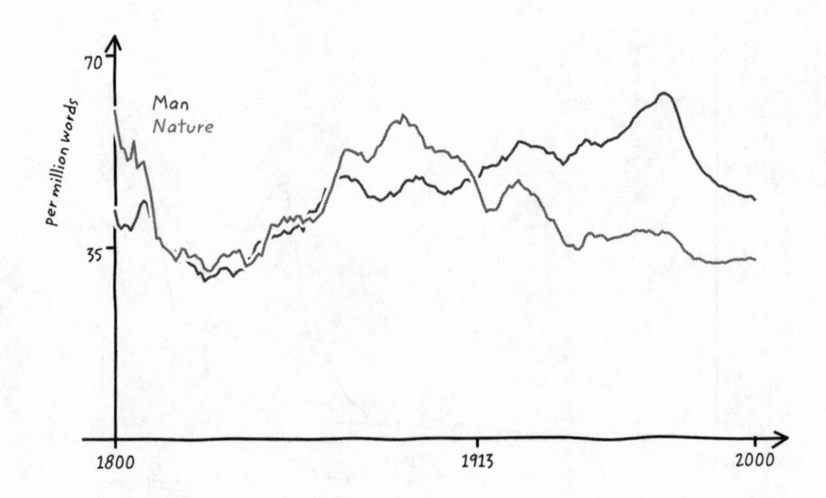

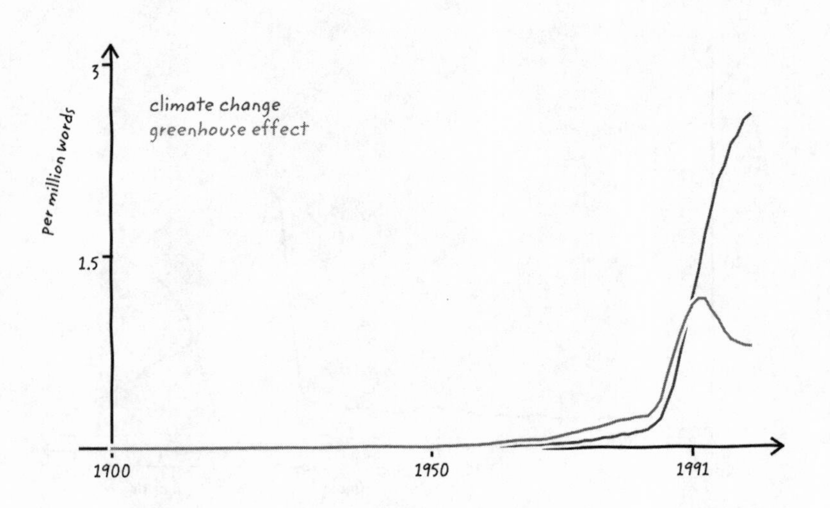

The World

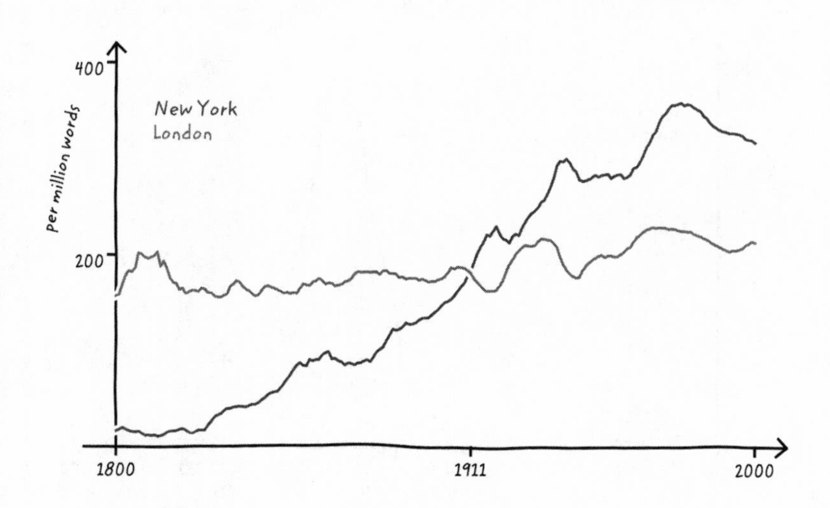
400
200
Per million words
New York
London
1800
1911
2000

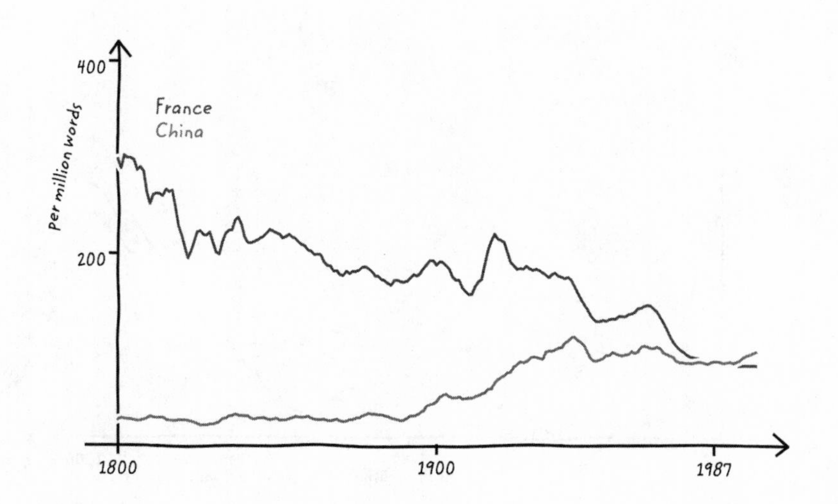
400
200
Per million words
France
China
1800
1900
1987

Engineering

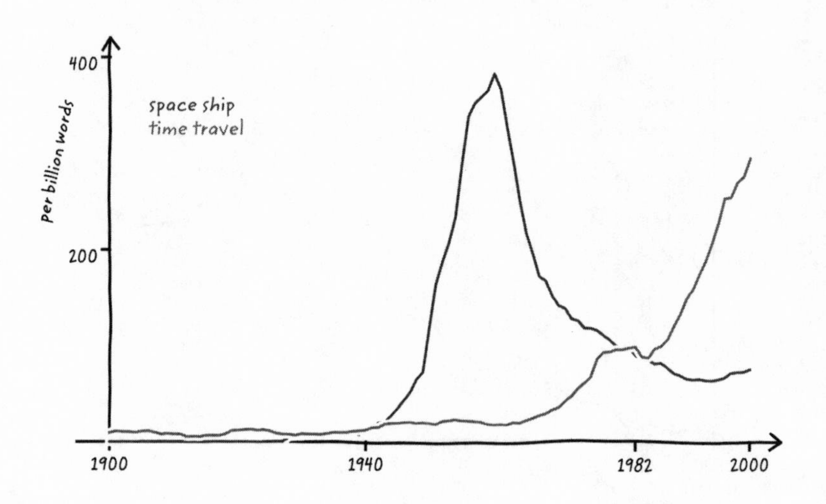

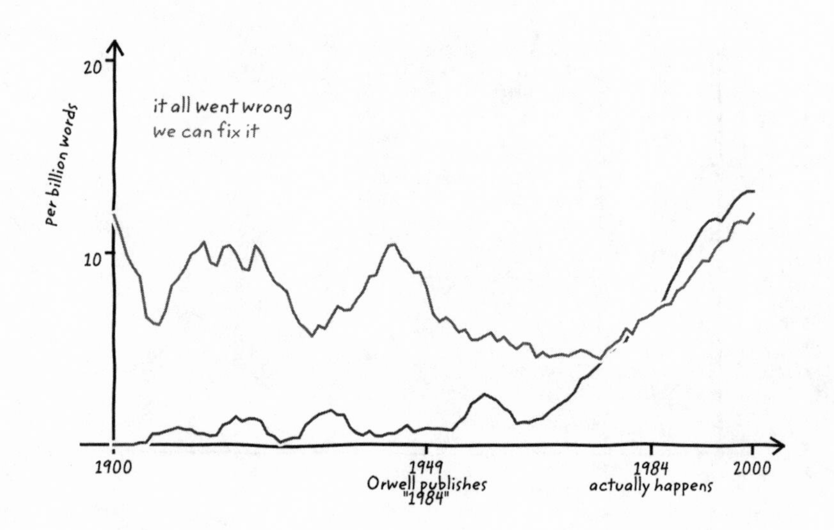

Fight Night

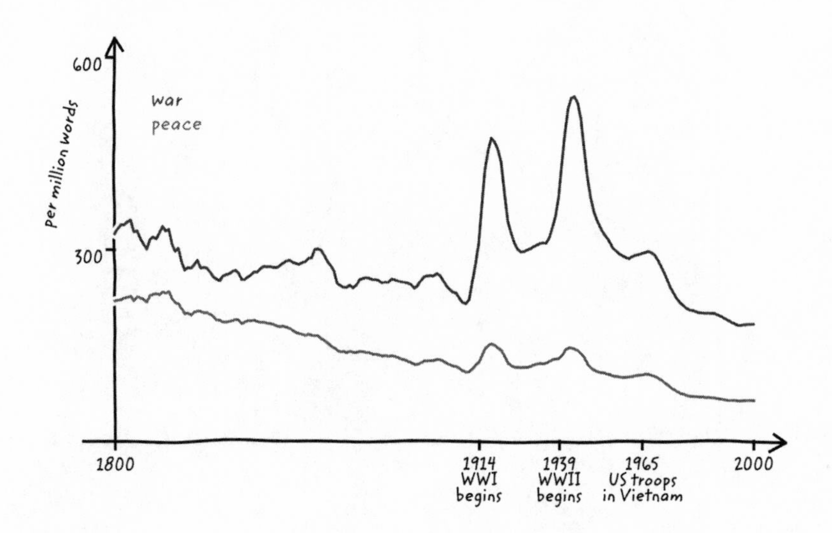
600
300
per million words
war
peace
1800
1914
WWI
begins
1939
WWII
begins
1965
US troops
in Vietnam
2000

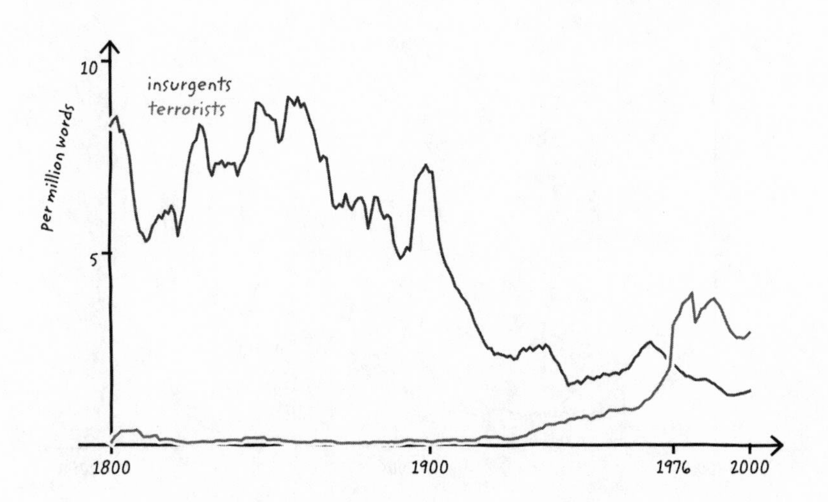
10
5
per million words
insurgents
terrorists
1800
1900
1976
2000

Disease

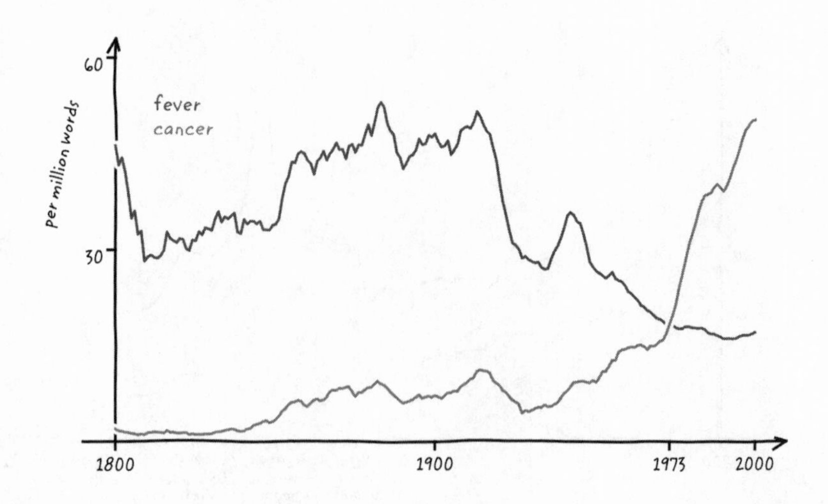
60
30
per million words
fever
cancer
1800
1900
1973
2000

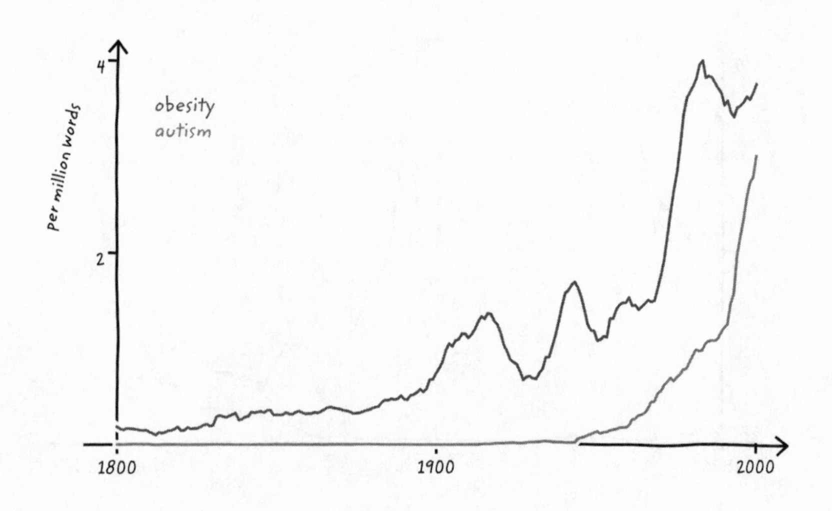
4
2
per million words
obesity
autism
1800
1900
2000

Medicine

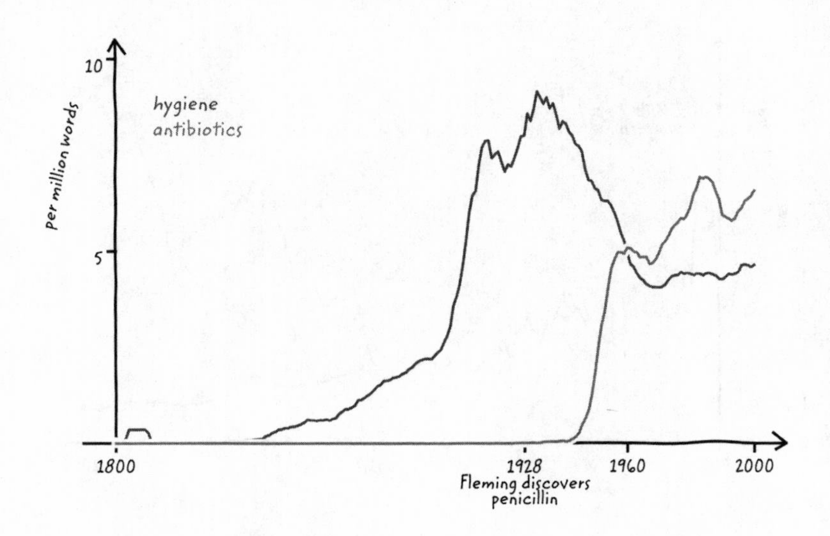
10
5
per million words
hygiene
antibiotics
1800
1928
1960
2000
Fleming discovers penicillin

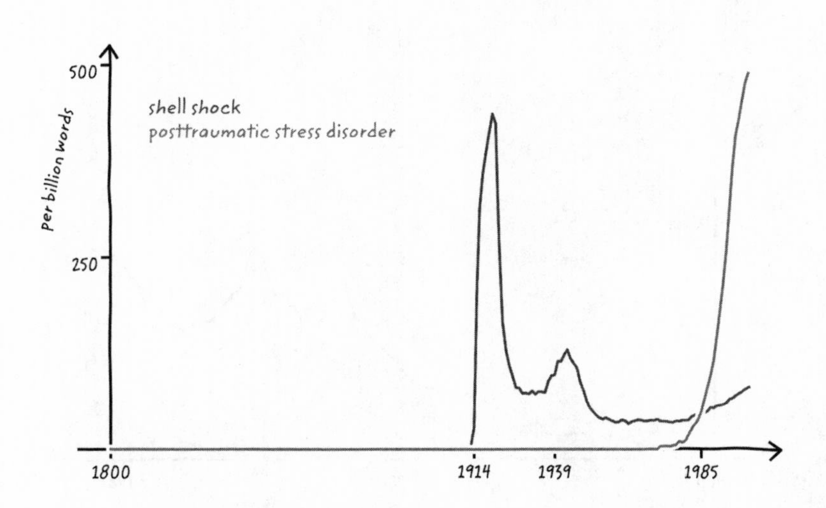
500
250
per billion words
shell shock
posttraumatic stress disorder
1800
1914
1939
1985

Eat Something

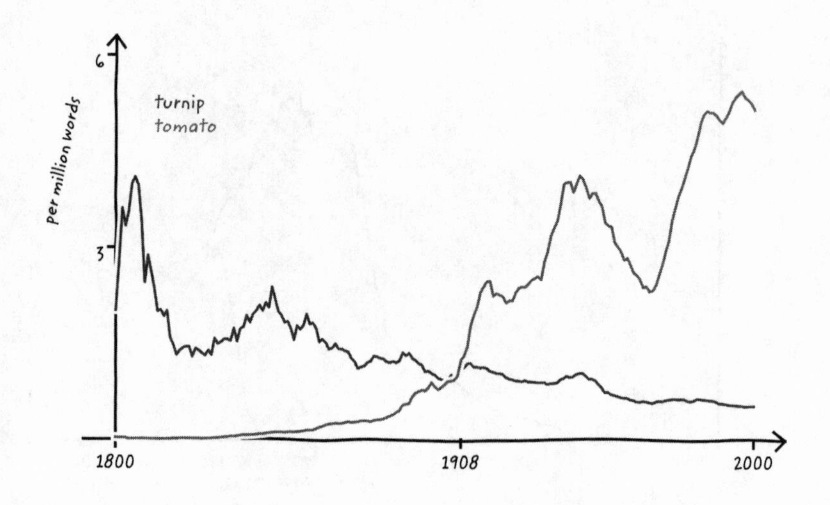

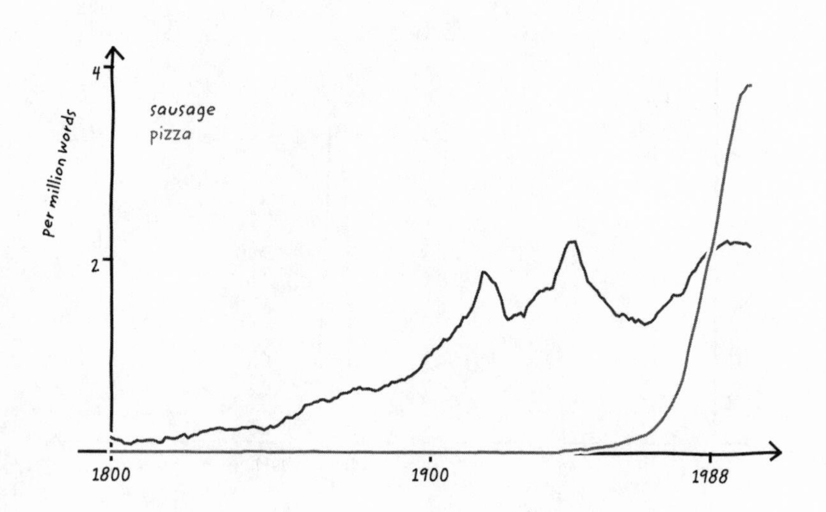

Drink Up

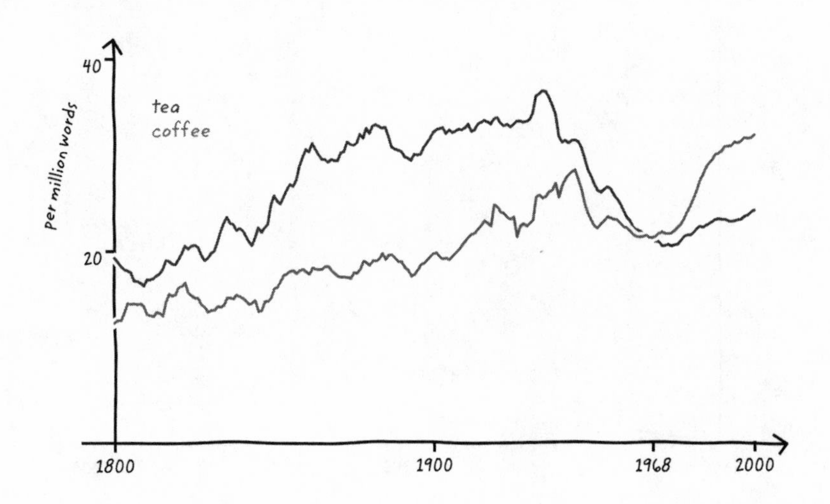
40
20
per million words
tea
coffee
1800
1900
1968
2000

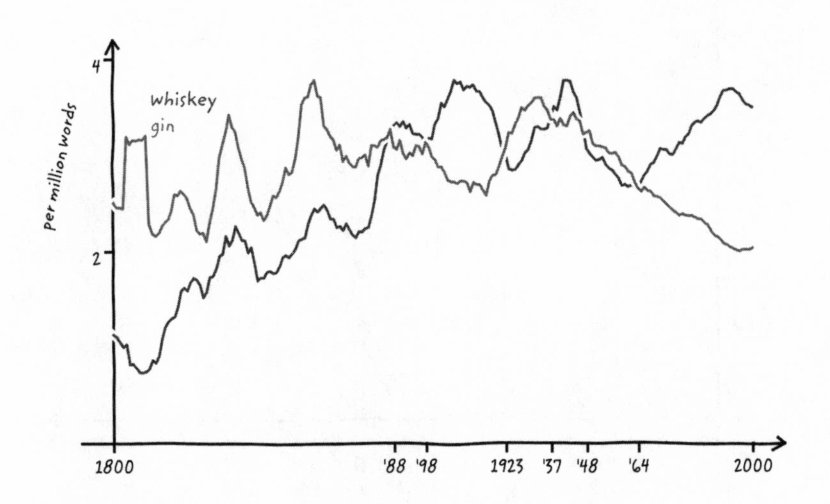
4
2
per million words
whiskey
gin
1800
'88
'98
1923
'37
'48
'64
2000

Fun and Games

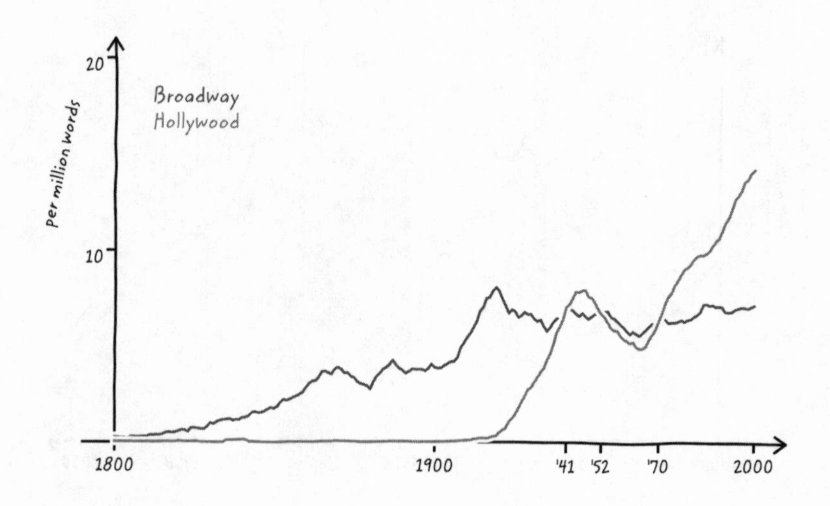

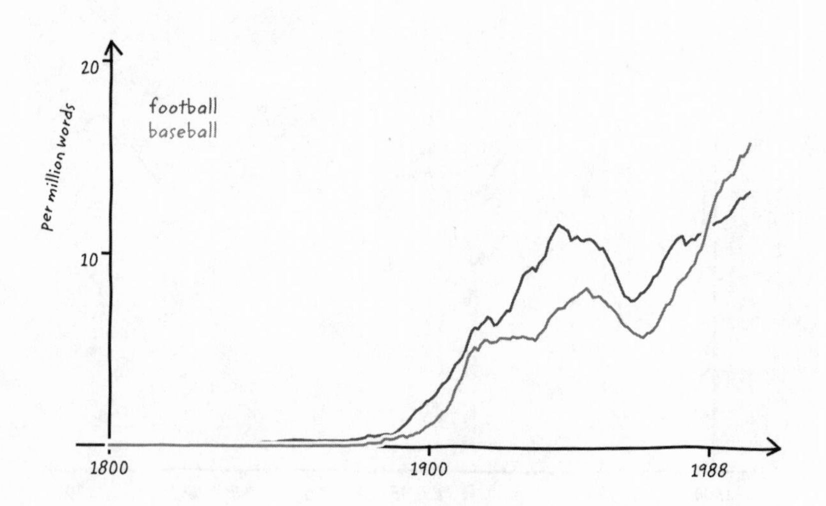

Nightlife

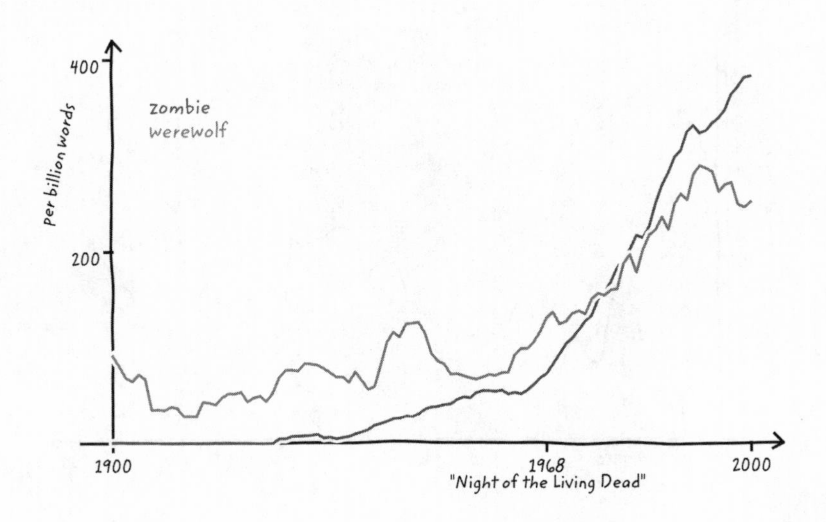

400
200
Per billion words
zombie
werewolf
1900
1968
"Night of the Living Dead"
2000

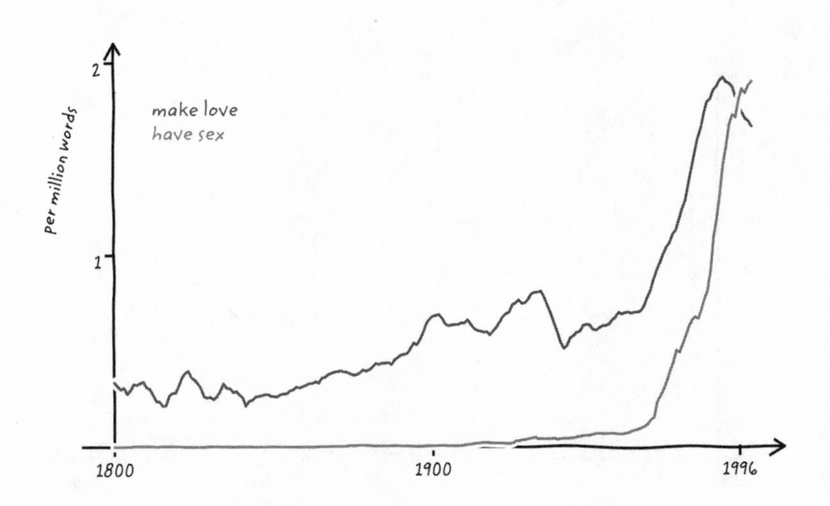

2
1
Per million words
make love
have sex
1800
1900
1996

Life Is Hard

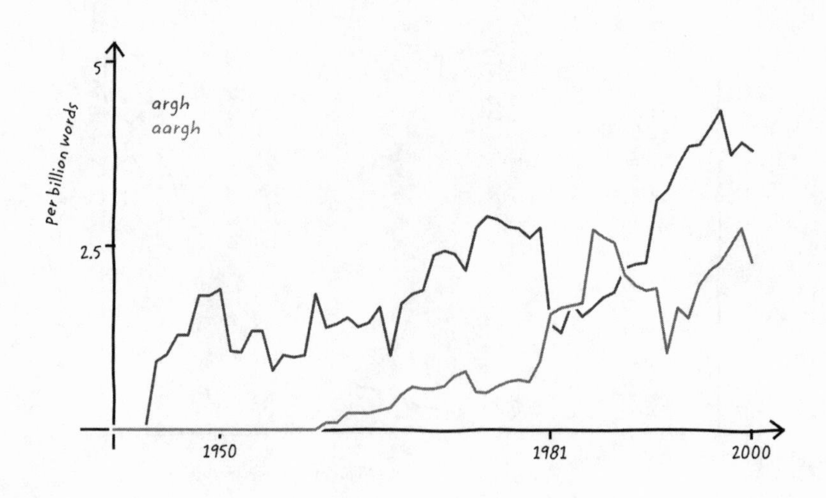

5
2.5
per billion words
argh
aargh
1950
1981
2000

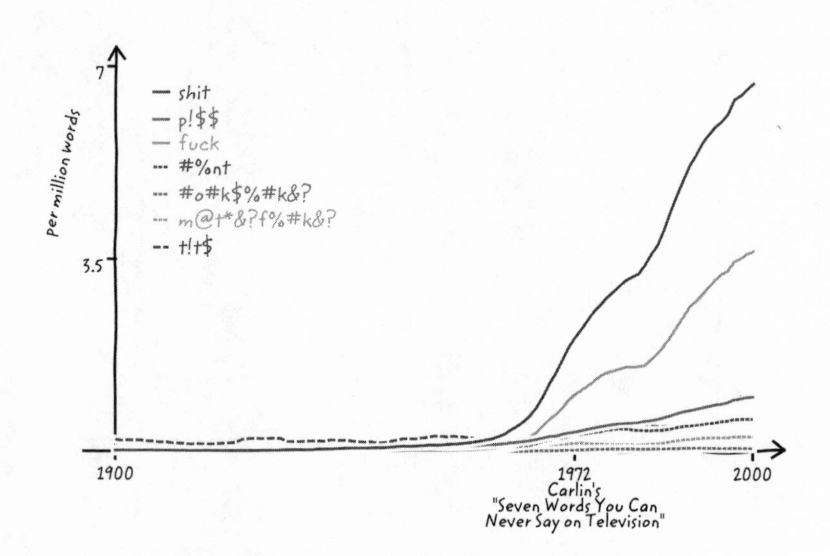

7
3.5
per million words
shit
p!$$
fuck
#%nt
#o#k$%#k&?
m@t*&?f%#k&?
t!t$
1900
1972
Carlin's "Seven Words You Can Never Say on Television"
2000

Modern Times

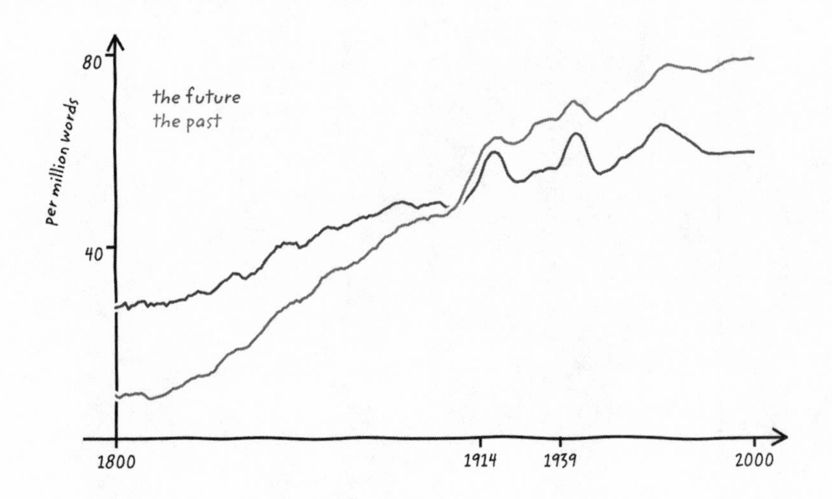

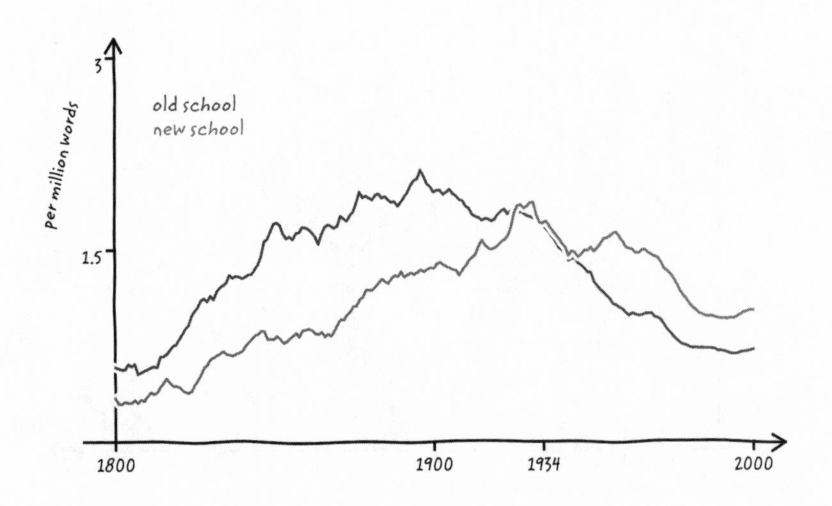

Great Minds

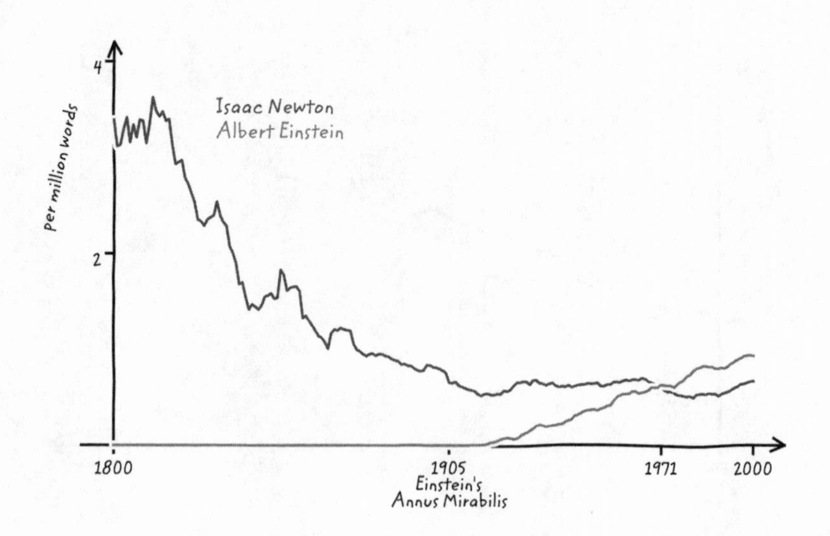
4
2
Per million words
Isaac Newton
Albert Einstein
1800
1905
Einstein's
Annus Mirabilis
1971
2000

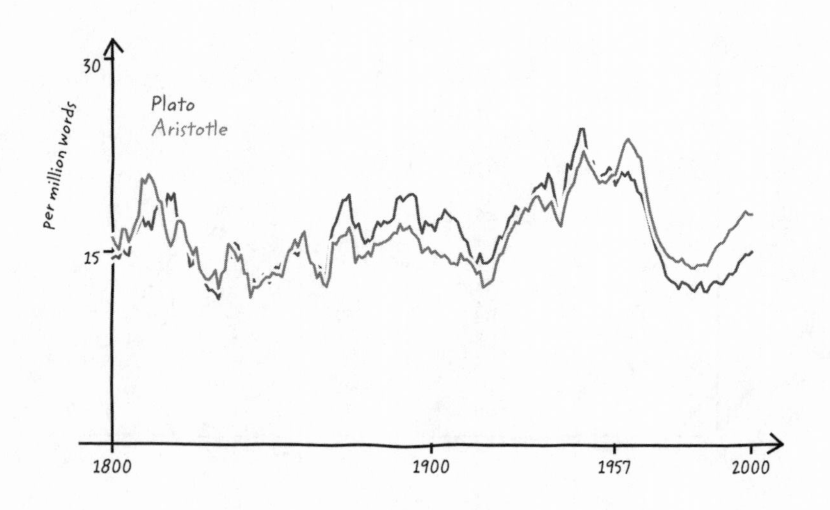
30
15
Per million words
Plato
Aristotle
1800
1900
1957
2000

Words of Wisdom

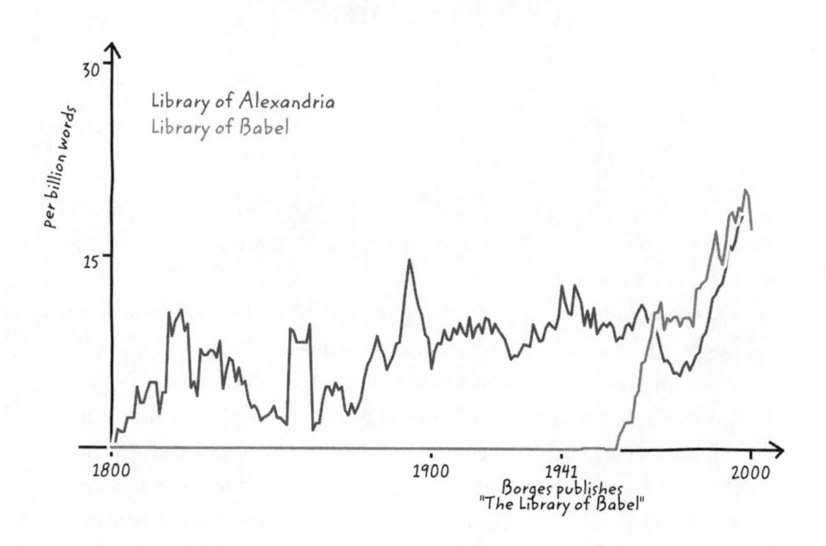

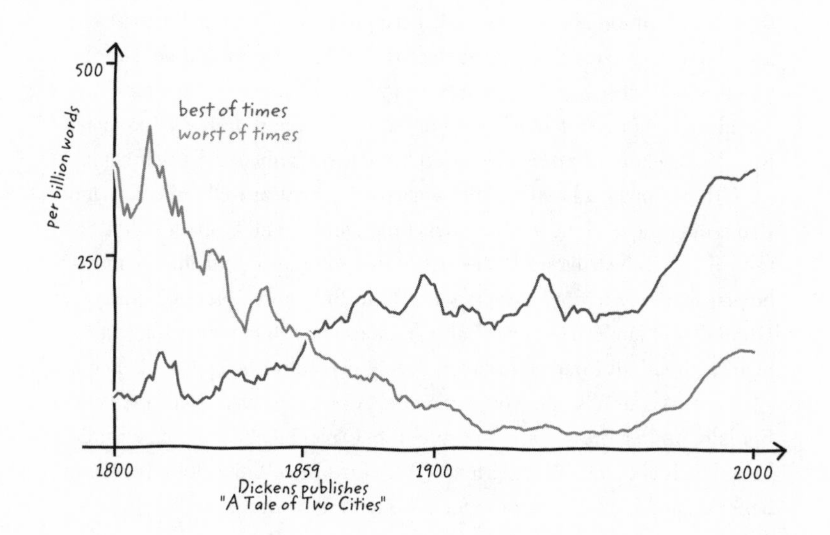

ACKNOWLEDGMENTS

At 2:00 p.m. on December 16, 2010, our article on culturomics appeared online, Google launched the Ngram Viewer, and the two of us breathed a sigh of relief: we were finally done!

Our break lasted until exactly 5:40 p.m., when Max Brockman—now our agent—sent us an e-mail titled, simply, "your book." Thank you, Max, for sending us that e-mail. Also: we are detail-obsessed perfectionists who, given a lily, immediately want to gild it, dip it in chocolate, and have it deep-fried. Thank you for going to bat, time and time again, for our crazy idea of a book.

This book would still be no more than a crazy idea if not for the extraordinary efforts of our editor, Laura Perciasepe, who fought extremely hard to make it a reality. She was a constant source of ideas, feedback, and reading assignments. It was always exciting to get a random package in the mail with some new paperback that she wanted us to read, an informal book-of-the-month club that shaped the present volume on the most fundamental level. We also owe a deep debt of gratitude to the designers and copy editors at Riverhead, whom we pestered to no end; and to our publicist, Katie Freeman.

Other people had a strong influence on this book as well. Julie Zauzmer stands out. She read the text countless times, and her ideas about everything from the overall structure of the text to the position of individual commas helped shape *Uncharted* at every scale. John Bohannon, Neva Cherniavsky Durand, and Jan Zauzmer were also gracious enough to review the text numerous times; all three contributed deeply perceptive insights and encouragement. We are also grateful to Samuel Arbesman, Ivan Bochkov, Pedro Bordalo, Andrea Bress, Elisheva Carlebach, Olga Dudchenko, Yitzie Ehrenberg, Sue Lieberman, Oliver Medvedik, Arina Omer, Suhas Rao, Benjamin Schmidt, and Elena Stamenova for commenting on drafts.

Science is a conversation. The ideas in this book are the fruits of a conversation that has involved too many wonderful collaborators to list. To prove it, here are but a few of their names: Aviva Aiden, Uri Alon, John Bohannon, Martin Camacho, Nicholas Christakis, Robert Darnton, Daniel Donoghue, Neva Cherniavsky Durand, Sara Eismann, George Fournier, Joseph Fruchter, Anthony Grafton, Jo Guldi, Joe Jackson, Eric Lander, Carol Lazell, Mark Liberman, Yuri Lin, Micheal Lopez, Sarah Johnson, Michael McCormick, Radhika Nagpal, Jeremy Rau, Charles Rosenberg, Tracey Robinson, Jonathan Saragosti, Benjamin Schmidt, Jesse Sheidlower, Yuan Shen, Stuart Shieber, Randy Stern, Tina Tang, Werner Treß, Adrian Veres, Ben Zimmer; Joe Pickett of the *American Heritage Dictionary*; Jorge Cauz, Carmen-Maria Hetrea, Dale Hoiberg, and Kunal Sen of the *Encyclopædia Britannica*; at Google, the entire Books team, notably Ben Bayer, Dan Bloomberg, Will Brockman, Ben Bunnell, Dan Clancy, Matt Gray, Peter Norvig, Jon Orwant, Slav Petrov, Ashok Popat, Leonid Taycher, Leslie Yeh, and especially Alfred Spector. Snippets of this conversation, highlighting many key contributors, appear throughout the text and notes; but for each anecdote we include, there are a half-dozen more we regret having had to leave out. Martin Nowak and Steven Pinker deserve to be singled out again: they have been essential catalysts of our work.

We, and our analyses, are little more than the sum of the books that we have read. We are grateful to everyone who has staked their name and reputation in this, most ancient art.

—Erez and JB

I am grateful to many people.

Helen Sultanik taught me science in sixth grade. Joel Wolowelsky taught me the elegance of mathematics. Dan Eshel let me play in his laboratory. The Johns Hopkins CTY program introduced me to other dorks. In these ways, I grew to love science.

Samuel Cohen, Robert Gunning, Saul Kripke, and Paul Seymour encouraged me as an undergraduate. Will Happer once said something he will not remember but that I will not forget—the best advice I ever got about how to choose a problem. Elisheva Carlebach, my master's thesis advisor, gave me a taste of the historian's life. Professor Martin Nowak took me under his wing, taught me how to write a scientific paper, taught me to embrace my humor, and believed in me. Eric Lander challenged me to be bold. Steven Pinker took our ideas seriously at a time when almost no one else did.

Lawrence David and Glen Weyl got saddled with a lot of yakking about culturomics, as did many other members of the endlessly interesting Harvard Society of Fellows.

Michael Berger, John Bohannon, Avi Bossewitch, Neal Dach, Sarah Johnson, Ari Packman, and Nicholas Christakis have really been there for me.

My lab: Ivan Bochkov, Martin Camacho, Ashok Cutkosky, Olga Dudchenko, Neva Cherniavsky Durand, Zach Frankel, Maxim Massenkoff, Matt Nicklay, Arina Omer, Suhas Rao, Adrian Sanborn, Benjamin Schmidt, Elena Stamenova, and Linfeng Yang; and the ROMEans who preceded them, all the way back to Joe Jackson, who spurred all of this with his undergraduate thesis. You guys make science fun.

Of course, no listing of scientific comrades-in-arms would be complete without my esteemed coauthor. JB, working with you for nearly a decade has been an incredible experience. So many discoveries, and so much fun along the way.

I am grateful for a loving family. My sisters, Tamar, Pattie, and Orly; their husbands, Ouri, David, and Eddie; and children, Ben, Danny, Eliana, Eshli, Gil, Isaac, Noah, Oren, and Zoë. Gil: thanks for all the proofreading over the years.

I am deeply thankful to my mother, Sue Lieberman. First, I exist. Second, she has always wanted the best for me, like when I was five years old and she tried to enroll me in Stuyvesant High School. Hang in there, Ima. I know it has been a rough ride of late.

I am thankful to my children, Gabriel Galileo, Maayan Amara, and Aragorn Banana, who fill my life with laughter and fun. My wife, Aviva, has been more wonderful than I could imagine for these eight years. She has been a rock while I wrote this book. She is brilliant, kind, patient, understanding, and supportive, a partner in science and, much more important, in life. It has been such a marvelous adventure.

I wish I could thank my father, Aharon Lieberman, עליו השלום. He passed away while we were writing this book, indeed while we were writing the chapter I most wanted to discuss with him. He was a genius, and a great inventor, and he gave me so many gifts, but none greater than his relentless enthusiasm for who I might become.

This book owes him a great debt. My father was, in his words, a "professional refugee," and he stomped around the English language without ever quite making it his home. When I was nine years old, he sat me down at a desk and made me write something for his company. I wrote it, very very badly. He couldn't have done it any better, but he knew that I, with the benefit of a native tongue, could. So he dissected my failure, in lengthy and unfaltering detail, and told me to do it again, from scratch. And again. And again. We worked in this way for eight years, on one project after another, until I left to go to college. And in this manner, the unlikeliest of teachers taught me how to write.

To my father, mediocrity was a moral failure. I miss him. I thank him. I pray he enjoys this book.

תם ונשלם שבח לאל ברוא עולם.

—Erez

Thank you, Ina: you are the most amazing person I know, my source of inspiration, strength, and happiness.

Most of who I am I owe to my close family: my father, Gilles, and my mother, Christine; my sister, Florence, and my brother, Marc-Antoine. I miss them, every day. My extended family plays no small role in my life: Mathieu and Thomas in Kuala Lumpur; my grandparents Mané and Daddy in Provence; all my family from Mauritius. I must have a special word for Grand-Mère, Dominique, Fabrice, Arnaud, Cédric, Cécile, Valérie, Aurélie, and Vanessa; and the Popovi and the Ivanovi in the UK and Bulgaria. They provide me with something truly invaluable: the emotional safety, the stability, and the happiness that make me free to take bolder risks.

I would be a profoundly different person today had I not crossed paths with Roy Kishony and Martin Nowak. They changed the course of my life when they exposed me to academic research. I can't thank them enough for giving me the space, support, guidance, and freedom to go to the boundaries of science and explore. I want to thank Steven Pinker again for being a wonderful mentor: his availability, thoughtfulness, and multiform support have helped me more than he knows.

I am grateful to Tim Mitchison for his unfailing scientific and moral support, for making Systems Biology a home for somewhat unconventional interests, and simply for being so forward-thinking. I am extremely grateful to Chris Sander and Debbie Marks for supporting me in many selfless ways, from providing space where I could write and do research, to entertaining endlessly creative conversations. I am deeply thankful to Mike McCormick, Bob Darnton, François Taddei, Tony Grafton, Franco Moretti, and Matthew Jockers for many enthralling conversations about science and the humanities.

It is worth repeating how grateful I am to the people who have directly contributed to culturomics, among them Aviva Aiden, John Bohannon, Martin Camacho, Neva Cherniavsky, Yuri Lin, Peter Norvig, Jon Orwant, Slav Petrov, Benjamin Schmidt, and Adrian Veres. This has been quite an adventure.

I thank Tom Rielly and Logan McClure for having gathered the most creative, fascinating, uplifting, yet unpretentious group of people I have ever met, the TED Fellows. These fellows are a continuous source of inspiration to me and to millions of others. On that note, I am grateful to members of DAMM at MEX for their creative energy and general cool.

Culturomics owes a great deal to the countless conversations I've had with my friends and colleagues over the years. Most of my ideas were first tested on Pedro Bordalo when they were still barely shadows of a mistaken thought. Thank you, my friend, for letting me shamelessly take advantage of your unbounded intellectual curiosity (still, let the record show that I own you at foosball—since it is written in a book, it must be true).

The following people helped develop culturomics by letting me bounce ideas at them, usually with their consent: Pamela Yeh, Tami Lieberman, and Remy Chait (thank you for helping me be a better collaborator); Kalin Vetsigian, Adam Palmer, Tobias Bollenbach, and Erdal Toprak (thank you for the little biology I know); Michael Manapat, Daniel Rosenbloom, Alison Hill, Tibor Antal, Anna Dreber, Thomas Pfeiffer, and Corina Tarnita (thank you for the math); and Fabien Azoulay, Marc Azoulay, Côme Denoyel, Neal Desai, Samuel Fraiberger, Bastien Guerin, Thomas Leonard, Nathan Leverence, Sidney Ouarzazi, Thibault Peyronel, Nick Stroustrup, and Mohamed Toumi.

And of course: thank you, Erez. This past decade exploring the frontiers of science with you was a fabulous experience; in so many ways, it shaped the person I am today.

—JB

NOTES

ABOUT THE CHARTS

The charts in this book were inspired by the delightful visual style of Randall Munroe's xkcd Web comic, http://xkcd.com/. The idea of automating xkcd-style graph production was proposed by Damon McDougall; the actual graphs in this book were created in Python, using a modified version of code by Jake VanderPlas. These ngrams can be generated, in interactive form, at the original Google Ngram Viewer, located at http://books.google.com/ngrams/, and, in xkcd style, at http://xkcd.culturomics.org. We hope Munroe won't mind. See http://xkcd.com/1007/ and http://xkcd.com/1140/. Some of his favorite ngrams appear at http://xkcd.com/ngram-charts/.

Note that ngram data is case-sensitive, and ngram plots depend on several parameters. Unless otherwise indicated in these notes, all ngram charts shown in the text correspond exactly to the results of the Google Ngram Viewer, using the English 2012 corpus and three-year smoothing. Unless otherwise noted, the query text is entirely in lowercase, except for proper nouns, which are capitalized in the usual way. All of the underlying datasets can also be downloaded at http://books.google.com/ngrams/datasets.

When referring to particular ngrams, such as *Marc Chagall* and *Kubismus* in the German corpus, we will cite them as NV: "Marc Chagall, Kubismus"/German. If no corpus is listed, the reference is to the English 2012 corpus, e.g., NV: "cubism." We sometimes note a year range or a smoothing value as well.

When using ngram data in a publication, please cite Jean-Baptiste Michel, Yuan Kui Shen, Aviva Presser Aiden, Adrian Veres, Matthew K. Gray, The Google Books Team, Joseph P. Pickett, Dale Hoiberg, Dan Clancy, Peter Norvig, Jon Orwant, Steven Pinker, Martin A. Nowak, and Erez Lieberman Aiden, "Quantitative Analysis of Culture Using Millions of Digitized Books," *Science* 331, no. 6014 (January 14, 2011; published online ahead of print December 16, 2010): 176–82.

CHAPTER 1. THROUGH THE LOOKING GLASS

Intro

2 **"the United States in *their* treaties with His Britannic Majesty."** Emphasis added.

2 **The Constitution.** The Constitution itself treats the United States as a plural. For instance, "Treason against the United States, shall consist only in levying War against them." See U.S. Const., art. III, §3.

2 **When the *is/are* switch took place.** The plural question clearly remained a live issue in 1901, when John W. Foster, who had served as secretary of state under President Benjamin Harrison, wrote an article debating the merits of the singular and plural forms in the *New York Times*. See John W. Foster, "Are or Is? Whether a Plural or a Singular Verb Goes with the Words United States," *New York Times*, online at http://goo.gl/Ql60b.

2 **". . . Certain large consequences of the war seem clear."** The quote is from James M. McPherson, *Battle Cry of Freedom* (Oxford: Oxford University Press, 1988), 859. We hope Professor McPherson does not mind, too much, our correcting an error in his deservedly celebrated work *Battle Cry of Freedom*. We highlighted it not as a criticism of his historical acumen, but precisely because McPherson, as historians go, is the best of the best. There is no better way to demonstrate the utility of these mechanical methods than by showing how even the greatest historians can use them.

2 **"There was a time a few years ago."** The quote is from the *Washington Post*, April 24, 1887, as quoted in Ben Zimmer, "Life in These, uh, This United States," *Language Log*, November 24, 2005, http://goo.gl/Ug8iX.

4 **Chart.** NV: "The United States is, The United States are." Note that without the initial

capital letter, one captures unwanted formulations, such as *The Senate of the United States is*, in which the *is* does not refer to *the United States* but to *The Senate of the United States*.

The Shape of the Light

5 **Lenses.** A richly detailed history of these developments appears in Vincent Ilardi, *Renaissance Vision from Spectacles to Telescopes* (Philadelphia: American Philosophical Society, 2007).

6 **Robert Hooke.** While writing this book, Erez visited Uppsala University in Sweden, where he had the opportunity to examine a 1665 first edition of Hooke's *Micrographia: or some physiological descriptions of minute bodies made by magnifying glasses with observations and inquiries thereupon*. Even by modern standards, Hooke's hand-drawn illustrations of what he saw through the microscope are spectacular. It is hard to imagine how visually stunning they would have been at the time. *Micrographia* was the first scientific bestseller, the ur-text of the popular science genre. Still, copies from the initial print run are very rare. Enter the digital book revolution: Today, anyone can peruse the original online. See Robert Hooke, *Micrographia* (London: Jo. Martyn and Ja. Allestry, 1665), online at http://goo.gl/KSnaH.

6 **Microbes.** First called animalcules by their discoverer, Antonie van Leeuwenhoek. See Clifford Dobell, *Antony van Leeuwenhoek and His "Little Animals"* (New York: Harcourt, Brace, 1932). Your own body contains ten times as many bacterial cells as human cells. See D. C. Savage, "Microbial Ecology of the Gastrointestinal Tract," *Annual Review of Microbiology* 31 (1977): 107, online at http://goo.gl/hzVlrR. The bacteria that live inside us outnumber the human population by a factor of about 10^{14}, or one hundred trillion.

6 **Magnification achieved by Galileo's telescope.** Galileo's very first telescopes were not as good; 30X was achieved only after several rounds of improvements. See Richard S. Westfall, "Science and Patronage: Galileo and the Telescope," *Isis* 76, no. 1 (March 1985): 11–30, online at http://goo.gl/eiPt3U; Henry C. King, *The History of the Telescope* (London: C. Griffin, 1955).

7 **Galileo's relationship to modernity.** See David Whitehouse, *Renaissance Genius: Galileo Galilei and His Legacy to Modern Science* (New York: Sterling, 2009); David Wootton, *Galileo: Watcher of the Skies* (New Haven, CT: Yale University Press, 2010); Mark Brake, *Revolution in Science: How Galileo and Darwin Changed Our World* (New York: Palgrave Macmillan, 2009); Jean Dietz Moss, *Novelties in the Heavens: Rhetoric and Science in the Copernican Controversy* (Chicago: University of Chicago Press, 1993); Robert S. Westman, *The Copernican Question: Prognostication, Skepticism, and Celestial Order* (Berkeley: University of California Press, 2011).

Counting Sheep

9 **The birth of writing.** The early history of human writing was uncovered in large part through the pioneering work of Denise Schmandt-Besserat. What Schmandt-Besserat has called "the Rosetta stone of the token system"—one of the most important finds in the archaeology of ancient writing—is a hollow tablet discovered at Nuzi in Iraq from the second millennium BCE. The cuneiform inscription on the outside of the tablet reads: "21 ewes that lambed//6 female lambs//8 full-grown male sheep//4 male lambs//6 she-goats that kid//1 he-goat//3 female kids//The Seal of Ziqarru, the shepherd." When the tablet was opened up, forty-nine counters were found inside: one for each animal listed on the outside. Why the redundancy? The inscription on the outside could be easily referred to, but also could be easily tampered with. The inside, though difficult to refer to, would be hard to tamper with. Thus, a dispute between the parties to the contract could be adjudicated by breaking open the tablet to reveal the counters inside. Scholars believe that, after some time, people realized that cuneiform could be used on the inside as well as the outside, eliminating the need for counters and making it possible to create legal documents that used writing alone. The

practice of creating contracts in which part of the writing was left "open" for easy reference, and part was sealed for the purpose of adjudicating disputes, became common; an example of this sort of contract appears in the Hebrew Bible, Jeremiah 32:10–11. See Barry B. Powell, *Writing: Theory and History of the Technology of Civilization* (Chichester, England: Wiley-Blackwell, 2009); Richard Rudgley, *The Lost Civilizations of the Stone Age* (New York: Free Press, 1999); Denise Schmandt-Besserat, *How Writing Came About* (Austin: University of Texas Press, 1996); Denise Schmandt-Besserat, *Before Writing*, vol. 1, *From Counting to Cuneiform* (Austin: University of Texas Press, 1992); Denise Schmandt-Besserat, *Before Writing*, vol. 2, *A Catalog of Near Eastern Tokens* (Austin: University of Texas Press, 1992). Of course, unanimity is rare among researchers. Some argue that writing emerged independently in Egypt, possibly via a quite different mechanism. See Larkin Mitchell, "Earliest Egyptian Glyphs," *Archaeology* 52, no. 2 (March/April 1999), online at http://goo.gl/tM3GEQ.

Big Data

11 **Bits and bytes.** The classic game twenty questions could also be called "two and a half bytes," because that's how much information you're allowed to collect before you guess.

11 **Five zettabytes.** Projection based on the IDC "Digital Universe" report. See John Gantz and David Reinsel, "The Digital Universe in 2020," EMC Corporation, December 2012, http://idcdocserv.com/1414. See also "Data, Data Everywhere," *Economist*, February 25, 2010, online at http://goo.gl/VsXh5P; Roger E. Bohn and James E. Short, "How Much Information? 2009," Global Information Industry Center, January 2010, http://goo.gl/pt0R; Peter Lyman and Hal R. Varian, "How Much Information? 2003," University of California at Berkeley, http://goo.gl/vpo9N.

11 **Writing out information.** We assume the typical bit takes six millimeters to write. This depends to some extent on the ratio of ones to zeros, since "1" is very narrow. Typical letter sizes for handwritten text are noted in Vikram Kamath et al., "Development of an Automated Handwriting Analysis System," *ARPN Journal of Engineering and Applied Sciences* 6, no. 9 (September 2011), online at http://goo.gl/4mlkTm.

11 **Sheep counting.** Thus, counting sheep is a completely solved problem, unless the universe expands very considerably.

11 **Doubling rate.** According to IDC estimates, humanity's data footprint will grow from 130 exabytes in 2005 up to 40,000 exabytes (40 zettabytes) in 2020. This suggests a doubling rate of about one year and ten months. See above.

The Digital Lens

12 **Size of Facebook.** See "Facebook Tops 1 Billion Users," Associated Press, October 4, 2012, online at http://goo.gl/nfK32P.

13 **Jon Levin.** See Liran Einav et al., "Learning from Seller Experiments in Online Markets," National Bureau of Economic Research (September 2011), online at http://goo.gl/f9ghir.

13 **James Fowler.** See Robert M. Bond et al., "A 61-Million-Person Experiment in Social Influence and Political Mobilization," *Nature* 489, no. 7415 (2012): 295–98, online at http://goo.gl/AQdAS0.

13 **Albert-László Barabási.** See Chaoming Song et al., "Limits of Predictability in Human Mobility," *Science* 327, no. 5968 (2010): 1018–21, online at http://goo.gl/rYlF2v.

14 **Jeremy Ginsberg.** See Jeremy Ginsberg et al., "Detecting Influenza Epidemics Using Search Engine Query Data," *Nature* 457 (2009): 1012–14, online at http://goo.gl/WHEWW.

14 **Raj Chetty.** See Raj Chetty, John N. Friedman, and Jonah E. Rockoff, "The Long-Term Impacts of Teachers," National Bureau of Economic Research (December 2011), online at http://goo.gl/C18JQ; Raj Chetty et al., "How Does Your Kindergarten Classroom Affect Your Earnings?," National Bureau of Economic Research (March 2011), online at http://goo.gl/N9O6a.

15 **Nate Silver.** See Nate Silver, FiveThirtyEight, http://www.fivethirtyeight.com; Nate Silver, *The Signal and the Noise* (New York: Penguin, 2012).

The Library of Everything

15 **Every book.** What does this actually mean? There's not much point in digitizing every copy of every book ever written—although we wouldn't say that there's no point at all; people's marginal notes can be fascinating. See Anthony Grafton and Joanna Weinberg, *I Have Always Loved the Holy Tongue* (Cambridge, MA: Harvard University Press, 2011). On the other hand, numerous editions of the most famous works can appear over the centuries and can differ substantially from one another. This can get pretty hairy. See, for instance, Eric Rumsey, "Google Book Search: Multiple Editions Give Quirky Results," Seeing the Picture, October 12, 2010, http://goo.gl/6YNld. In the case of Google Books, the goal is to digitize one copy of every book edition.

16 **Stanford Digital Library Technologies Project.** See "The Stanford Digital Library Technologies Project," Stanford University, http://goo.gl/tstLQ; "Google Books History," Google Books, http://goo.gl/ueobb.

16 **Size of Google Books.** Partly for the reasons pointed out above, and partly because the definition of a book, as a physical object, is ambiguous, counting the number of books in a physical library also can be tricky. As such, the number of books in each library was obtained from the library's page on Wikipedia on July 18, 2013. Note that these numbers are not equally up-to-date. Also note that Stanford is already beginning to close physical libraries and replace them with "bookless libraries." See Lisa M. Krieger, "Stanford University Prepares for the 'Bookless Library,'" *San Jose Mercury News*, May 18, 2010, online at http://goo.gl/yauezp.

Long Data

17 **The books we checked.** For instance, see the digitized edition of Louis F. Klipstein, *Grammar of the Anglo-Saxon Language* (New York: George P. Putnam, 1848), online at http://goo.gl/cWRlJ. Note that, in light of legal and ethical concerns, Harvard ended up opting out of the Google Books program, allowing Google to digitize only out-of-copyright works. See Laura G. Mirviss, "Harvard-Google Online Book Deal at Risk," *Harvard Crimson*, October 30, 2008, online at http://goo.gl/0tYflD.

18 **Long data.** This term was recently coined by social network researcher Samuel Arbesman. See Samuel Arbesman, "Stop Hyping Big Data and Start Paying Attention to 'Long Data,'" *Wired*, January 29, 2013, http://goo.gl/X7oEC.

Mo' Data, Mo' Problems

21 **Problems sharing data.** Despite the fact that the best empirical datasets are not broadly available, social networks remain a rich area for research. See, for instance, Duncan J. Watts and Steven H. Strogatz, "Collective Dynamics of 'Small-World' Networks," *Nature* 393, no. 6684 (1998): 440–42, online at http://goo.gl/be3Xmi; Albert-László Barabási and Réka Albert, "Emergence of Scaling in Random Networks," *Science* 286, no. 5439 (1999): 509–12, online at http://goo.gl/eESUa8; Ron Milo et al., "Network Motifs: Simple Building Blocks of Complex Networks," *Science* 298, no. 5594 (2002): 824–27, online at http://goo.gl/duzS5L.

22 **Lawyers.** Note that sometimes, lawyers can be a good omen. One of us is married to a lawyer.

Culturomics

22 **Launch of culturomics.** We initially released four resources summarizing our findings: a scientific paper, a detailed methodological supplement, and two supplemental Web sites. See Jean-Baptiste Michel et al., "Quantitative Analysis of Culture Using Millions of Digitized Books," *Science* 331, no. 6014 (January 14, 2011), online at http://goo.gl/mahoN; extensive supplemental text, online at http://goo.gl/1e509; "Ngram Viewer," Google Books,

2010, http://books.google.com/ngrams; "Culturomics," Cultural Observatory, http://www.culturomics.org. Because we will refer to Michel et al. frequently in these notes, we will abbreviate the reference as Michel2011. We will use Michel2011S to refer to the paper's supplemental text.

23 **Our new scope.** See "Ngram Viewer," above; Erez Lieberman Aiden and Jean-Baptiste Michel, "Culturomics, Ngrams and New Power Tools for Science," Google Research Blog, August 10, 2011, http://goo.gl/FSbbP; Jon Orwant, "Ngram Viewer 2.0," Google Research Blog, October 18, 2012, http://goo.gl/zOSfg.

How many words is a picture worth?

24 **Brisbane's speech to marketers.** In 1911, extracts from his talk to the Syracuse, New York, Advertising Club appeared in *Printers' Ink*, the first American trade publication for the advertising industry. These extracts contain the earliest recorded form of the expression: "Use a picture. It's worth a thousand words." The more compact form, "A picture is worth a thousand words," appears shortly thereafter, as do the "ten thousand" and "million" variants; initially, all three versions are typically attributed to Brisbane. It's quite possible that he said all three in different contexts. See *Printers' Ink* 75, no. 1 (April 6, 1911): 17. By 1925, the phrase was being attributed directly to Confucius.

See also *Management Accounting*, National Association of Cost Accountants (1925).

CHAPTER 2. G. K. ZIPF AND THE FOSSIL HUNTERS

Intro

26 **"beautiful beautiful beautiful beautiful beautiful."** See Karen Reimer, *Legendary, Lexical, Loquacious Love* (Chicago: Sara Ranchouse, 1996).

26 **Karen Reimer.** More precisely, the book's cover attributes the book to "Karen Reimer writing as Eve Rhymer." For more info about Karen Reimer's work, see http://www.karenreimer.info.

Problem Child

28 **Big data.** The big data trend is a bit too recent to be easily seen in books; see our discussion of the time resolution of the book record in chapter 6. Other big data will have to suffice. According to Google Trends, the search volume for *big data* at Google was relatively flat until 2011, and then began to surge. The Wikipedia article on "Big Data" was created in April 2010; as of July 14, 2013, it has been edited 694 times, is viewed more than 150,000 times a month, and is the 2,022nd most popular article on the English Wikipedia. See: "Big data," Google Trends, 2013, http://goo.gl/tL8GnD; "Big Data," Wikipedia, July 14, 2013, http://goo.gl/DFFbr; "Big Data: Revision History," Wikipedia, July 14, 2013, http://goo.gl/Jvla3; "Big Data," X!'s Edit Counter, July 14, 2013, http://goo.gl/e9YZ7v; "Big Data," Wikipedia Article Traffic Statistics, July 14, 2013, http://goo.gl/vgYxH.

28 **Program for Evolutionary Dynamics.** There's no better way to get a sense of the place, the research, and the man in charge than through Nowak's book on the topic. See Martin A. Nowak with Roger Highfield, *SuperCooperators* (New York: Free Press, 2011).

29 **Where the sun goes at night.** The answer is discussed in a controversial work, originally published in 1632, by Galileo Galilei. See his *Dialogue Concerning the Two Chief World Systems, Ptolemaic and Copernican*, trans. Stillman Drake (New York: Modern Library, 2001).

29 **Why the sky is blue.** The effect is due to Rayleigh scattering, discovered by Lord Rayleigh. At the time, his name was John Strutt. See John Strutt, "On the Light from the Sky, Its Polarization and Colour," *Philosophical Magazine* 41, series 4 (1871): 107–20, 274–79.

29 **Whether a tree could grow as tall as a mountain.** See George W. Koch et al., "The Limits to Tree Height," *Nature* 428 (April 22, 2004): 851–54, online at http://goo.gl/lxNlq.

29 **Why you have to go to sleep.** See Carlos Schenck, *Sleep* (New York: Penguin, 2007). Despite the existence of numerous books on the subject, nobody really knows why we need to sleep. It's a fun area for theorists. See, for instance, Van M. Savage and Geoffrey B. West, "A Quantitative, Theoretical Framework for Understanding Mammalian Sleep," *PNAS: Proceedings of the National Academy of Sciences* (November 20, 2006), online at http://goo.gl/wFWDC.

Dinosaur Hunters

31 **Anthropology as science.** See Nicholas Wade, "Anthropology a Science? Statement Deepens a Rift," *New York Times*, December 9, 2010, online at http://goo.gl/eCI9K3.

31 **Nathan Myhrvold.** See Nathan Myhrvold, Chris Young, and Maxine Bilet, *Modernist Cuisine: The Art and Science of Cooking* (Bellevue, WA: The Cooking Lab, 2011); Malcolm Gladwell, "In the Air," *New Yorker*, May 12, 2008, online at http://goo.gl/TTtsLU.

1937: A Data Odyssey

32 **Frequency of *the*.** Frequency in English books in 2000: 4.6 per hundred words.

32 **Frequency of *quiescence*.** Frequency in English books in 2000: two in every five million words.

33 **Counting words today.** The following Linux command produces a list of all 1-grams in a text file, ordered from most frequent to least frequent:

cat textfile.txt | tr ' ' '\n' | sort | uniq -c | sort -k1 -n -r > 1grams.txt

33 **Human computers.** Many of these human computers were women. Their remarkable story is told in David Alan Grier, *When Computers Were Human* (Princeton, NJ: Princeton University Press, 2007). Amazon's Mechanical Turk service, billed as "artificial artificial intelligence," reflects in some ways a Web-based, crowdsourced return to this sort of approach. See http://www.mturk.com.

33 **Miles L. Hanley.** See Miles Hanley, *Word Index to James Joyce's* Ulysses (Madison: University of Wisconsin Press, 1937).

33 **Role of Hanley's word index in Zipf's work.** Zipf's first encounter with the law that bears his name precedes his examination of word frequency in *Ulysses*. In 1911, a businessman named R. C. Eldridge published a list of word frequencies calculated using eight pages of newspaper text. Having noticed "that a moderate number of words, wisely selected, would enable any two people understanding them . . . to converse intelligently on many subjects," Eldridge's goal was to use lexical statistics to outline "the foundations of a universal vocabulary." The resulting frequencies were the basis of Zipf's calculations in his 1935 book *Psycho-Biology of Language*, which is the first of Zipf's publications on the regularity now known as Zipf's law. See George Kingsley Zipf, *The Psycho-Biology of Language* (Boston: Houghton Mifflin, 1935), online at http://goo.gl/KYvOcK; George Kingsley Zipf, *Human Behavior and the Principle of Least Effort* (Reading, MA: Addison-Wesley, 1949); R. C. Eldridge, *Six Thousand Common English Words* (Buffalo, NY: Clement Press, 1911).

33 **Ranking the words in Ulysses by frequency.** Zipf was able to rely extensively on an appendix to Hanley's word index, by Martin Joos, in which Joos tabulated most of the requisite statistics.

34 **Zipf's law.** We'd be remiss if we failed to point out that Zipf's law is neither Zipf's, nor is it a law. It's not a law for several reasons. First, it's only approximately true; on close examination, most languages exhibit systematic deviations away from purely Zipfian behavior. Second, despite many (conflicting) theoretical derivations, it's not clear that Zipf's law must hold for all languages, or for any language in particular. Zipf's law is best thought of as an extremely universal—and rather mysterious—empirical regularity.

It's also not really Zipf's, because Zipf was not the first to discover it. As far as we know, the first person to uncover the underlying mathematical principle was a French stenographer named Jean-Baptiste Estoup, who began publishing his explorations on this topic in the 1912 edition of his popular work on shorthand note-taking, a discipline in which Zipfian regularities have immediate and practical consequences. The classic representation of Zipf's law by means of a rank-frequency plot on double-log axes was first introduced by Edward Condon in a 1928 paper in *Science*. Condon went on to become a very prominent physicist, serving as president of both the American Physical Society and the American Association for the Advancement of Science.

Zipf's first publication on Zipf's law appeared in 1935. He appears to have independently rediscovered many of the findings of the others, and confirmed them using much better data. (A critical examination of Zipf's intellectual debts, although fascinating, is beyond the scope of this text.) Zipf continued to work on the subject for many years, setting the basic results in the context of both a theoretical framework and a broad examination of similar phenomena throughout the social sciences. Zipf also served as the single most influential synthesizer and popularizer of these ideas. A review of his 1949 book *Human Behavior and the Principle of Least Effort* called it "one of the most ambitious books ever written . . . altogether different and refreshing. It cuts across departmental and divisional boundaries as nothing else has for a century." See John Q. Stewart, review of *Human Behavior and the Principle of Least Effort*, by George Kingsley Zipf, *Science* 110, no. 2868 (December 16, 1949): 669. For the sake of conciseness, our discussion in the main text is loosely based on the treatment given in this book.

Still, given the fuller history of this concept, is there a more accurate name for Zipf's law? It's pretty reasonable to argue that Zipf's law really ought to be called the Estoup-Condon-Zipf regularity. Even that's not totally fair. Zipf's work was made possible by the word indexing and counting that had been performed by Hanley, Joos, and Eldridge. Condon's work, too, was based on frequency analyses performed by others—in his case, Leonard Ayres and Godfrey Dewey (son of Melvil Dewey, who invented the Dewey decimal system). So really we should call Zipf's law the Estoup-Condon-Zipf-Eldridge-Ayres-Dewey-Hanley-Joos regularity. This is probably why we just stick to Zipf's law.

Anyway, it's almost a truism that every finding based on painstaking analysis of a really impressive dataset is not named after the person who generated the underlying data. While we're busy naming things, we might as well hand out consolation prizes. Call this one the Hanley principle.

See Jean-Baptiste Estoup, *Gammes Sténographiques* (Paris: Institut Sténographique, 1916); E. U. Condon, "Statistics of Vocabulary," *Science* 67, no. 1733 (March 16, 1928): 300, online at http://goo.gl/Qi5B49; Leonard P. Ayres, *A Measuring Scale for Ability in Spelling* (New York: Russell Sage Foundation, 1915), online at http://goo.gl/C0cgke; Godfrey Dewey, *Relative Frequency of English Speech Sounds* (Cambridge, MA: Harvard University Press, 1923); M. Petruszewycz, "L'Histoire de la Loi d'Estoup-Zipf: Documents," *Mathématiques et Sciences Humaines* 44 (1973): 41–56, online at http://goo.gl/LlrNn.

A brief and elegant review of these ideas appears in Willem Levelt, *A History of Psycholinguistics* (Oxford: Oxford University Press, 2012). A very extensive bibliography on Zipf's law and related principles is given in Nelson H. F. Beebe, *A Bibliography of Publications about Benford's Law, Heaps' Law, and Zipf's Law* (Salt Lake City: University of Utah, 2013), online at http://goo.gl/TuyT0. A related concept is the notion of "1/f noise." See Benoit B. Mandelbrot, *Multifractals and 1/f Noise: Wild Self-Affinity in Physics* (New York: Springer, 1999).

The World According to Zipf

34 **Distribution of human height.** See C. D. Fryar, Q. Gu, and C. L. Ogden, "Anthropometric Reference Data for Children and Adults: United States, 2007–2010," *Vital Health Statistics* 11, no. 252 (2012), online at http://goo.gl/uEuiV.

35 **Power laws.** More precisely, something is said to be a power law when one quantity is proportional to another quantity, elevated to a fixed exponent, or power. Zipf's law is a power

law in which the two quantities are rank and abundance, and the exponent equals one. If the quantities pertain to a network, the underlying network is in general known as "scale-free." See Steven H. Strogatz, "Exploring Complex Networks," *Nature* 410, no. 6825 (2001): 268–76, online at http://goo.gl/gO6Eb4. When the two quantities pertain to a geometric structure, and the exponent is not an integer, there is a special word for the underlying structure: a fractal. See Benoit Mandelbrot, *The Fractal Geometry of Nature* (San Francisco: W. H. Freeman, 1985).

Although Zipf was among the first to identify a power law in word frequencies, earlier researchers had discovered other power laws in entirely different disciplines. Most notably, the economist Vilfredo Pareto observed that 80 percent of the land in Italy was owned by 20 percent of the people. This was the first of many such 80/20 rules. This sort of skew is, mathematically speaking, closely associated with power laws.

Many power-law relationships were first reported by Zipf in Zipf, 1949, where he also collects many findings by others. For more recent surveys, see Aaron Clauset, Cosma Rohilla Shalizi, and M. E. J. Newman, "Power-Law Distributions in Empirical Data," *SIAM Review* 51, no. 4 (2009): 661–703, online at http://goo.gl/6PLJFF; Manfred Schroeder, *Fractals, Chaos, Power Laws: Minutes from an Infinite Paradise* (New York: W. H. Freeman, 1991). Such relations are so ubiquitous that there can be a vast array of examples in seemingly narrow fields. See, for instance, Ignacio Rodríguez-Iturbe and Andrea Rinaldo, *Fractal River Basins: Chance and Self-Organization* (Cambridge, England: Cambridge University Press, 2001).

36 **Bill Gates vs. the moon.** According to the 2010 census, the median net worth of American households, excluding home equity, was $15,000. In March 2010, *Forbes* estimated Bill Gates' net worth at $53 billion. Five-foot-seven is 1.7 meters. Thus in our hypothetical scenario, Gates would be about 6,007 kilometers tall. This is far taller than Pluto (diameter 2,390 kilometers), Mercury (diameter 4,879 kilometers), and the moon (diameter 3,474 kilometers); it's nearly as tall as Mars (diameter 6,792 kilometers). Even if home equity is included, bringing the median household net worth up to $66,740, he would still be 1,350 kilometers tall, more than half the height of Pluto. See "The World's Billionaires: William Gates III," *Forbes*, March 10, 2010, http://goo.gl/8ykj; "Wealth and Asset Ownership," U.S. Census Bureau, July 11, 2013, http://goo.gl/llnbC and in particular "Wealth Tables 2010," U.S. Census Bureau, http://goo.gl/v7mxk.

36 **Reasons behind Zipf's law.** See M. E. J. Newman, "Power Laws, Pareto Distributions and Zipf's Law," *Contemporary Physics* 46, issue 5 (2005), online at http://goo.gl/nrkMB. The random monkeys explanation appears in George A. Miller, "Some Effects of Intermittent Silence," *American Journal of Psychology* 70, no. 2 (June 1957): 311–14, online at http://goo.gl/p6PLll.

Too Zipf or Not Too Zipf

37 **The irregular verbs.** For a rich and detailed introduction to this fascinating topic, see Steven Pinker, *Words and Rules: The Ingredients of Language* (New York: Basic Books, 1999). Depending on your perspective, irregular verbs are either strange or delightfully quirky. A woman once ran a personals ad in the *New York Review of Books* that began: "Are you an irregular verb?" See Steven Pinker, *The Language Instinct* (New York: William Morrow, 1994), 134.

The Few, the Proud, the Strong

38 **Learning irregular verbs.** Children master irregular verbs in a particularly fascinating way, going through characteristic stages that correspond to their increasingly sophisticated minds. At first, they conjugate all verbs idiosyncratically. Then they begin to recognize the rules inherent in the language spoken around them. When they realize that most verbs obey the *-ed* rule, they pass into a stage called hyperregularization, in which they treat every verb as regular and say things like *goed* and *knowed* and *runned*. Eventually, they realize that

certain verbs are exceptions to the *-ed* rule and gradually begin to incorporate the correct irregular forms into their speech.

39 **Proto-Indo-European and the ablaut.** See J. P. Mallory and D. Q. Adams, *The Oxford Introduction to Proto-Indo-European and the Proto-Indo-European World* (Oxford: Oxford University Press, 2006); Don Ringe, *A Linguistic History of English* (Oxford: Oxford University Press, 2006).

39 **The emergence of the dental suffix.** Unlike the strong irregulars, regulars are also known as "weak." See Detlef Stark, *The Old English Weak Verbs* (Tübingen, Germany: M. Niemeyer, 1982); Robert Howren, "The Generation of Old English Weak Verbs," *Language* 43, no. 3 (September 1967), online at http://goo.gl/2yf0t.

40 **Irregularization.** Regularization is usually a one-way street, but there are extremely rare exceptions. One is the irregular form *snuck*, which sneaked into the English language this past century. Following the lead of irregular verbs like *stick/stuck, strike/struck*, and *stink/stunk*, about 1 percent of English speakers are switching from *sneaked* to *snuck* each year. At this rate, one person will have snuck off while you read this sentence. See Steven Pinker, "The Irregular Verbs," *Landfall* (Autumn 2000): 83–85, online at http://goo.gl/kFFzLm.

2005: Another Data Odyssey

41 **Why we say *drove*.** Actually, there's no such thing as a completely irregular verb in Modern English. Even if its frequency is very low, the regular form always exists, biding its time. Frequency has a very strong effect on this phenomenon, because frequent irregulars do a much better job of suppressing the competing regular form. Compared to *drove*, the signal for *drived* is negligible. That probably keeps *drove* safe. In contrast, *throve* has been looking vulnerable for centuries; the regularized form *thrived* started winning in the twentieth century but was already a formidable competitor long before that. This is a very general effect. In our ngram data, we found *found* (frequency: 1 in 2,000) 200,000 times more often than we finded *finded*. But *dwelt* (frequency: 1 in 100,000) dwelt in our data only 60 times as often as *dwelled* dwelled. See Michel2011.

Note that, for the purpose of our 2007 study, we occasionally needed a list of Modern English irregular verbs that we could regard as "authoritative." For instance, we used such a list to determine which verbs have regularized and which have not. Curating such a list on our own could leave our method vulnerable to concerns about cherry-picking, so we used the list that appears in S. Pinker and A. Prince, "On Language and Connectionism: Analysis of a Parallel Distributed Processing Model of Language Acquisition," *Cognition* 28 (1988): 73–193. We regarded as irregular any verb that has at least one sense which is conjugated as an irregular according to this list. Note that there is occasional disagreement between dictionaries and other sources about which verbs are irregular and which are not. For instance, *wed/wed* remains irregular according to the above list, but not according to all contemporary dictionaries. (Some already favor *wed/wedded*.)

41 **Extant linguistic hypotheses.** The relationship between frequency and regularization is explored in Joan L. Bybee, *Morphology: A Study of the Relation Between Meaning and Form* (Amsterdam: John Benjamins, 1985). More generally, there has been a great deal of work on how linguistic change comes about. See, for instance, William Labov, "Transmission and Diffusion," *Language* 83, no. 2 (June 2007): 344–87, online at http://goo.gl/aZ5M2R; Greville Corbett et al., "Frequency, Regularity, and the Paradigm: A Perspective from Russian on a Complex Relation," in *Frequency and the Emergence of Linguistic Structure*, ed. Joan L. Bybee and Paul J. Hopper (Amsterdam: John Benjamins, 2001), 201–28. These questions can also be explored from a more explicitly evolutionary perspective. See Mark Pagel, *Wired for Culture: Origins of the Human Social Mind* (New York: W. W. Norton, 2012); Mark Pagel, Quentin D. Atkinson, and Andrew Meade, "Frequency of Word-Use Predicts Rates of Lexical Evolution Throughout Indo-European History," *Nature* 449 (October 11, 2007): 717–20, online at http://goo.gl/93WiJ0. See also Partha Niyogi, *The Computational Nature*

of Language Learning and Evolution (Cambridge, MA: MIT Press, 2009); Niyogi, a luminary in the field, passed away, tragically, in 2010. He was forty-three.

42 **Textbooks.** These include, for instance, Oliver Farrar Emerson, *A Middle English Reader* (New York: Macmillan, 1909), and Henry Sweet, *An Anglo-Saxon Primer* (Oxford: Clarendon Press, 1887).

Survival of the Fit

43 **Detecting natural selection.** There is a massive literature on this topic. See, e.g., P. C. Sabeti et al., "Detecting Recent Positive Selection in the Human Genome from Haplotype Structure," *Nature* 419, no. 6909 (2002): 832–37, online at http://goo.gl/TW6SYJ; P. Varilly et al., "Genome-Wide Detection and Characterization of Positive Selection in Human Populations," *Nature* 449, no. 7164 (2007): 913–18, online at http://goo.gl/NfnzeU.

44 **Our analysis of irregular verbs.** This work originally appeared as Erez Lieberman et al., "Quantifying the Evolutionary Dynamics of Language," *Nature* 449 (October 11, 2007): 713–16, online at http://goo.gl/3kCMQT.

44 **Radioactivity and half-life.** See "Radioactive Decay," Wikipedia, June 22, 2013, http://goo.gl/xTYh1; "Half-life," Wikipedia, June 3, 2013, http://goo.gl/TXn3.

The Once and Future Past

46 **When *drove* will regularize.** The half-life of irregular verbs as frequent as *drove* is 5,400 years, which is equivalent to an expected lifetime of about 7,800 years before it regularizes.

John Harvard's Shiny Shoe

46 **Visitors rub the shoe with their hands.** But it's not just hands that keep the shoe shiny. Many undergrads also urinate on the shoe; in 2013, 23 percent of graduating Harvard seniors reported having done so. Putting the "john" back in John Harvard is one of the "big three" rites of passage for Harvard undergraduates. The second is a nude yelling ritual known as primal scream. The third is having sex in Widener Library, demonstrating the student body's continuing enthusiasm for getting physical with books. Try doing that with a Kindle. See Julie M. Zauzmer, "Where We Stand: The Class of 2013 Senior Survey," *Harvard Crimson*, May 28, 2013, online at http://goo.gl/1EpfA.

Lexicon and Concord

48 **Creating concordances.** Some concordances are more powerful than others. It must be pointed out that, even if you set aside the much more challenging source material, Busa's concordance was far more sophisticated than Reimer's. For instance, the *Index Thomisticus* incorporates a complete lemmatization of the underlying text, grouping all words into lexically related classes. (In English, a lemmatization would group words like *run, running, runs, ran, outrun*, and *also-ran* under a single heading.) This lemmatization is itself a remarkable accomplishment. The ngram datasets we released do not feature lemmatization. It's very hard to do well.

48 **The *Index Thomisticus*.** In 1980, Busa published a firsthand account of his decades-long collaboration with IBM. It is an astonishingly prescient document, packed with too many insights to enumerate. For instance, anticipating the need for big humanities (see also our discussion in chapter 7), Busa writes:

> *Today's academic life seems to be more in favor of many short-term research projects which need to be published quickly, rather than of projects requiring teams of coworkers collaborating for decades. . . . It would be much better to build up results one centimetre at a time on a base one kilometre wide, than to build up a kilometre of research on a one centimetre base.*

More than thirty years later, Anthony Grafton, then president of the American Historical Association, expressed a similar train of thought:

> *As new forms of scientific research offer historians research possibilities that complement the textual record, as digital archives and exhibitions expand and digital research methods become more accessible, historians will have to learn how to form and work in teams. . . . Collaboration offers one way—potentially a very powerful one—for scholars of traditional bent to create global histories of economic, cultural, and political relations that rest on deep archival and textual foundations.*

Arguably the founding document of the digital humanities movement, Busa's account remains required reading to this day. See R. Busa, "The Annals of Humanities Computing: The *Index Thomisticus*," *Computers and the Humanities* 14 (1980): 83–90, online at http://goo.gl/FgVWQ; A. Grafton, "Loneliness and Freedom," *Perspectives on History*, March 2011, online at http://goo.gl/dOx3J.

Taking Roses Apart to Count Their Petals

49 **"Take roses apart."** See G. A. Miller, introduction to *The Psycho-Biology of Language* (Cambridge, MA: MIT Press, 1965), online at http://goo.gl/KYvOcK. The full quote, from the very beginning of his 1965 introduction, is as relevant today as it ever was:

> The Psycho-Biology of Language *is not calculated to please every taste. Zipf was the kind of man who would take roses apart to count their petals; if it violates your sense of values to tabulate the different words in a Shakespearean sonnet, this is not a book for you. Zipf took a scientist's view of language—and for him that meant the statistical analysis of language as a biological, psychological, social process. If such analysis repels you, then leave your language alone and avoid George Kingsley Zipf like the plague. You will be much happier reading Mark Twain: "There are liars, damned liars, and statisticians." Or W. H. Auden: "Thou shalt not sit with statisticians nor commit a social science."*
>
> *However, for those who do not flinch to see beauty murdered in a good cause, Zipf's scientific exertions yielded some wonderfully unexpected results to boggle the mind and tease the imagination.*

Burnt, baby, burnt

52 **Michael Phelps.** See Sally Jenkins, "Burned-Out Phelps Fizzles in the Water against Lochte," *Washington Post*, July 29, 2012.

52 **Kobe Bryant.** See Melissa Rohlin, "Kobe Bryant Says He Learned a Lot from Phil Jackson," *Los Angeles Times*, November 14, 2012, online at http://goo.gl/bKGDTg.

52 **An alliance of irregular verbs.** See the discussion of this topic in Steven Pinker, *Words and Rules: The Ingredients of Language* (New York: Basic Books, 1999); Lieberman et al., "Quantifying the Evolutionary Dynamics of Language," and its supplemental materials; Michel2011 and Michel2011S.

52 **Cambridge, England.** Here we assume that the *burned*-to-*burnt* frequency ratio reflects the proportion of English speakers in the United Kingdom who use each form.

CHAPTER 3. ARMCHAIR LEXICOGRAPHEROLOGISTS

Intro

53 ***Sasquatch.*** See Jeff Meldrum, *Sasquatch: Legend Meets Science* (New York: Forge, 2006).

54 ***Chupacabra.*** These creatures, and many more, are discussed in Loren Coleman and Jerome Clark, *Cryptozoology A to Z* (New York: Fireside, 1999). Note that chupacabras travel in packs; if you happen to run into one in a sentence, there's a decent chance that there are others lurking nearby. The frequency of *chupacabra* is surging right now, so they will probably be more common in the future.

Twenty-nine-year-old Billionaire Psychology

55 **Google Books.** See "Google Books History," http://goo.gl/ueobb.

55 **Digitization project time estimates.** Five hundred years for the University of Michigan is just multiplication; Coleman's estimate of a thousand years presumably includes time for doing things other than flipping pages, and of course may not have assumed just one person doing the flipping. Assuming 130 million books and forty minutes per book, it would take 9,900 years to do them all.

Page's Pages

56 **Lexical bulletin.** It seems possible to construct an English sentence of arbitrary length consisting only of the words *page* and *pages*. For instance:

"Page!" (Marissa Mayer, commanding someone to turn pages.)

"Page, page!" (Marissa, commanding Larry.)

"Page, page pages!" (A more detailed instruction.)

"Page, page Page's pages!" (By paging someone else's pages, Larry was screwing things up.)

"Page, page Page's page's pages." (Page's page was falling behind.)

"Page, page pages Page's page pages." (Marissa, telling a page to page the pages that the specific page assigned to Larry usually pages.)

56 **130 million books.** See Leonid Taycher, "Books of the world, stand up and be counted! All 129,864,880 of you," Google Books Search, August 5, 2010, http://goo.gl/5yNV. Taycher is Google's chief metadata guru.

56 **Nondestructive scanning.** As anyone who's ever tried to Xerox a book page would know, getting good images can be tricky. Here's just one of the many problems that needed to be overcome: Pages in books don't like to lie flat; they curve inward as one gets close to the binding. To solve this problem, Google developed a system for correcting each image to account for the curvature of the page. A much more extensive discussion of this process appears in Michel2011S.

57 **Gallup.** Gallup's seven-day averages were based on surveys of approximately 2,700 likely voters. See "Election 2012 Likely Voters Trial Heat: Obama vs. Romney," Gallup, http://goo.gl/ujbzb.

Twenty-five-year-old Graduate Student Psychology

58 **Peter Norvig.** For his MOOC, see "Introduction to Artificial Intelligence," https://www.ai-class.com/. For his textbook, see Stuart J. Russell and Peter Norvig, *Artificial Intelligence: A Modern Approach* (Englewood Cliffs, NJ: Prentice Hall, 1995).

Fortune 500 Legal Department Psychology

59 **Legal issues.** Wikipedia has been keeping close track of the lawsuits and their complex, ongoing development. See "Google Book Search Settlement," Wikipedia, June 23, 2013, http://goo.gl/8E5Cx. Some of the legal issues are discussed in Giovanna Occhipinti Trigona, "Google Book Search Choices," *Journal of Intellectual Property Law and Practice* 6, no. 4 (March 10, 2011): 262–73, and more generally in Marshall A. Leaffer, *Understanding Copyright Law*, 5th ed. (Albany, NY: Matthew Bender, 2011). A very detailed bibliography on this topic is kept at Charles W. Bailey, Jr., "Google Books Bibliography," *Digital Scholarship*, 2011, http://goo.gl/grff2. See Rubin's remarks at Thomas C. Rubin, "Searching for Principles: Online Services and Intellectual Property," Microsoft, http://goo.gl/GX3CB.

Big Data Casts Big Shadows

61 **America Online.** See Michael Barbaro and Tom Zeller, Jr., "A Face Is Exposed for AOL Searcher No. 4417749," *New York Times*, August 9, 2006, http://goo.gl/c8MCY; "About AOL Search Data Scandal," http://goo.gl/6hnfuI.

In the Shadow of Google Books

65 **The basis of modern genome sequencing.** Because of its relevance to genome sequencing, an extensive theoretical apparatus already exists for analyzing the problem of how well you can assemble whole texts from tiny text-tiles. The watershed moment in this literature was the development of the Lander-Waterman statistics. Because of dramatic improvements in genome sequencing technology, and due to the complex repeat structure of mammalian genomes, these statistics actually apply at least as readily to ngram-based attacks on whole text corpora as they do to the output of contemporary genome sequencers. See E. S. Lander and M. S. Waterman, "Genomic Mapping by Fingerprinting Random Clones," *Genomics* 2, no. 3 (April 1988): 231–39, online at http://goo.gl/wuAcXr.

Leaders of the Free Word

67 **"Potatoe."** See Dan Quayle, *Standing Firm* (New York: HarperCollins, 1994); Mark Fass, "How Do You Spell Regret? One Man's Take on It," *New York Times*, August 29, 2004, online at http://goo.gl/gWW4wK.

67 ***Refudiated.*** Palin famously used the 1-gram in a tweet on July 18, 2010. She had previously used the word on television. See Max Read, "Sarah Palin Invents New Word: 'Refudiate,'" *Gawker*, July 19, 2010, online at http://goo.gl/XjV7TJ.

67 **"Shakespeare liked to coin new words too".** See Michael Macrone, *Brush Up Your Shakespeare* (New York: HarperCollins, 1990); Jeffrey McQuain and Stanley Malless, *Coined by Shakespeare* (Springfield, MA: Merriam-Webster, 1998).

To Word, or Not to Word?

68 ***American Heritage Dictionary.*** Despite its linguistically conservative reputation, *AHD* has long been, from the methodological standpoint, extremely innovative.

In 1967, Henry Kucera and W. Nelson Francis published the "Brown Corpus," a million-word text collection meant to be representative of a broad array of genres. This publication proved instrumental in the emergence of corpus linguistics as an academic discipline, and is therefore, in many ways, the earliest and most important forerunner of the corpus we created at Google.

Shortly thereafter, publisher Houghton Mifflin approached Kucera about creating a corpus to assist with the new dictionary that the company was creating. Essentially, the publisher intended to put the strategy of Eldridge (see the notes to chapter 2's "1937: A Data Odyssey") into practice, using lexical statistics to construct a vocabulary of the English language. The first edition of Houghton Mifflin's *American Heritage Dictionary*, which appeared in 1969, was the first dictionary to employ such a strategy.

It was therefore natural to wonder how well the trailblazing *AHD* might hold up in light of our powerful new Google Books–based corpus. Luckily, Joseph P. Pickett, who was executive editor of the *AHD* from 1997 to 2011, was happy to participate. Thus, all of our analyses of the *American Heritage Dictionary* benefited immensely from his active collaboration, as well as from the assistance of his staff. All of the numbers reported about the *AHD* in this book are based on communication with Pickett and his team, as well as data that they provided. (Pickett was ultimately a coauthor of Michel2011.) Although we do critique the *AHD* at times in the text, it was clear the *AHD* felt that aggressively pursuing new types of analysis could help make the best possible dictionary. We think transparency in linguistic governance is a great idea, and no other reference work proved as transparent as the *AHD*.

The *AHD* famously relies on a usage panel. This panel consists of about two hundred language experts from all walks of life, ranging from Supreme Court justice Antonin Scalia to *New York Times* crossword editor Will Shortz to Pulitzer Prize–winning author Junot Díaz. It is chaired by Steven Pinker (also a coauthor of Michel2011). The panel represents, in many ways, the opposite of the culturomics or text-corpus-statistics approach to tracking language. It doesn't rely on representative sampling of language use in general, but instead on a small number of language experts—a lexical elite.

We wondered how these two approaches would compare. Each year, the *AHD* sends out a questionnaire to its usage panel. One year, the *AHD* allowed us to create our own supplement to this questionnaire, which the panelists also filled out. We compared the results to our ngram findings. For example, we asked them about *sneaked* and *snuck*: Which of these past-tense forms did the panelists find acceptable? We found that younger panelists were more likely to find *snuck* acceptable (unpublished data). Our ngram findings show the rapid spread of *snuck* in the last few decades. Taken together, these results may suggest that panelists, and perhaps language users more generally, tend to form their notions of what is or is not acceptable usage at a young age.

See *American Heritage Dictionary of the English Language*, 4th ed. (Boston: Houghton Mifflin, 2000); "The Usage Panel," *American Heritage Dictionary*, 2013, http://goo.gl/JtT4l; Francis Nelson and Henry Kucera, *Brown Corpus Manual* (Brown University Department of Linguistics, 1979).

68 **Number of words in the *AHD*.** The *AHD* team provided us with a list of the 153,459 headwords of all entries in the fourth edition of their dictionary. Sometimes, the same word appeared multiple times on the list; for instance, *console* appeared as both a noun and a verb. We removed such multiples. We also removed headwords that were not single words, like *men's room*. The resulting word list contained 116,156 words.

68 **Number of words in the *OED*.** This number is for the *OED*'s last printed edition, the second edition of 1989. (Many, including the CEO of Oxford University Press, Nigel Portwood, suspect the third edition will never appear in print, because of the general migration of such references to the Web.) Alas, we did not have the benefit of *OED* assistance. The *OED* Web site reports that the "number of word forms defined and/or illustrated" is 615,100. According to the preface, this edition also contained 169,000 "italicized-bold phrases and combinations," which are not 1-grams. Our estimate, 446,000, is just the difference between these two values. It is not an exact estimate, but rather an upper bound—the *OED*'s second edition does not have more 1-gram words than this value, but may have less. The *OED* recently invited us to be delegates to a symposium on its future, so perhaps a more robust, *AHD*-style collaboration is in the cards. It sure would be nice to get exact numbers. See *Oxford English Dictionary*, 2nd ed. (Oxford: Oxford University Press, 1989); "Dictionary Facts," *Oxford English Dictionary*, http://goo.gl/DL6a7; Bas Aarts and April McMahon, *The Handbook of English Linguistics* (Hoboken, NJ: John Wiley & Sons, 2008); Alastair Jamieson, "Oxford English Dictionary 'Will Not Be Printed Again,'" *Telegraph*, August 29, 2010, online at http://goo.gl/V5g8Ak.

69 **Prescriptive vs. descriptive.** See the ferocious public debates at Joan Acocella, "The English Wars," *New Yorker*, May 14, 2012, online at http://goo.gl/wGVHsx; Ryan Bloom, "Inescapably, You're Judged by Your Language," *New Yorker*, May 29, 2012, online at http://goo.gl/js9VJc; Steven Pinker, "False Fronts in the Language Wars," *Slate*, May 31, 2012, online at http://goo.gl/33vNYT. The debate also rages in academic circles. See, for instance, Henning Bergenholtz and Rufus H. Gouws, "A Functional Approach to the Choice Between Descriptive, Prescriptive and Proscriptive Lexicography," *Lexicos* 20 (2010), online at http://goo.gl/agXm7S.

69 **"Bull Moose" lexicography.** Roosevelt was supporting a plan first proposed by a group known as the Simplified Spelling Board. See David Wolman, *Righting the Mother Tongue: From Olde English to Email, the Tangled Story of English Spelling* (New York: Harper Perennial, 2010). An original letter of Roosevelt's on the topic can be seen, in digital facsimile, at "Letter from Theodore Roosevelt to William Dean Howells," Theodore Roosevelt Center at Dickinson State University, http://goo.gl/JA8cP.

69 **#ROFL.** Rolling on the floor laughing. If you don't know this, don't worry: Most dictionaries don't, either.

71 **Analysis.** All the analyses presented in the balance of the chapter are detailed in Michel2011 and Michel2011S.

DIY Dictionary

71 **Cutoff frequency.** We calculated the frequency distribution of the 116,156 unique 1-gram headwords in the *American Heritage Dictionary*. After the tenth percentile, at roughly one part per billion, the frequencies begin to soar.

72 **Words with nonalphabetical characters.** It's not at all clear that a word has to be composed entirely of alphabetic characters. For instance, the *OED* recently added, for the first time, an entry for a symbol, ♥. See Erica Ho, "The Oxford-English Dictionary Adds '♥' and 'LOL' as Words," *Time*, March 25, 2011, online at http://goo.gl/0RB6EA.

72 **Creating a Zipfian lexicon.** Note that this Zipfian lexicon is just a contemporary update of the idea, espoused by Eldridge and embodied in the *AHD*, that lexical statistics could be used to compile better dictionaries. An early, forceful argument to this effect appears in Richard W. Bailey, "Research Dictionaries," *American Speech* 44, no. 3 (1969): 166–72, online at http://goo.gl/4RqfDu.

Lexical Dark Matter

74 **Excluded categories.** Our choice of categories to be excluded (nonalphabetic terms, compounds that are easily understood from their component words, variant spellings, and undefinable terms) was based on discussions with Joseph Pickett of the *American Heritage Dictionary*. Standards vary somewhat from one to another, but broadly speaking, dictionaries have been deliberately excluding words for as long as they've been deliberately including them. Samuel Johnson discusses many examples of excluded words in his landmark 1755 dictionary. Dr. Johnson's ever-colorful ruminations on this topic in the dictionary's preface don't discuss the case of nonalphabetic terms, but do address the challenge of the other three classes.

Compounds, which he mostly left out: "*Compounded or double words I have seldom noted, except when they obtain a signification different from that which the components have in their simple state. Thus highwayman, woodman, and horsecourser, require an explication; but of thieflike or coachdriver no notice was needed, because the primitives contain the meaning of the compounds.*"

Variant spellings, which he mostly left in: "*I have not rejected any by design, merely because they were unnecessary or exuberant; but have received those which by different writers have been differently formed, as viscid, and viscidity, viscous, and viscosity.*" Spelling was much less standardized at the time.

Hard-to-define terms, in: "*Other words there are, of which the sense is too subtle and evanescent to be fixed in a paraphrase; such are all those which are by the grammarians termed expletives, and, in dead languages, are suffered to pass for empty sounds, of no other use than to fill a verse, or to modulate a period, but which are easily perceived in living tongues to have power and emphasis, though it be sometimes such as no other form of expression can convey.*"

He excludes many other categories as well, many of which remain common exclusion targets today.

Names: "*As my design was a dictionary, common or appellative, I have omitted all words which have relation to proper names; such as Arian, Socinian, Calvinist, Benedictine, Mahometan; but have retained those of a more general nature, as Heathen, Pagan.*" Jargon: "*That many terms of art and manufacture are omitted, must be frankly acknowledged; but for this defect I may boldly allege that it was unavoidable: I could not visit caverns to learn the miner's language, nor take a voyage to perfect my skill in the dialect of navigation, nor visit the warehouses of merchants, and shops of artificers, to gain the names of wares, tools and operations, of which no mention is found in books; what favourable accident, or easy enquiry brought within my reach, has not been neglected; but it had been a hopeless labour to glean up words, by courting living information, and contesting with the sullenness of one, and the roughness of another.*" In our analyses, Merriam-Webster's online dictionary often outperforms the OED on medical jargon, because the latter includes a separate, vast dictionary of medical terms (unpublished data). Foreign words: "*The words which our authours have introduced by their knowledge of foreign languages, or ignorance of their own, by vanity or wantonness, by compliance with fashion, or lust of innovation, I have registred as they occurred, though commonly only to censure them, and warn others against the folly of naturalizing useless foreigners to the injury of the natives.*" Fads: "*Nor are all words which are not found in the vocabulary, to be lamented as omissions. Of the laborious and mercantile part of the people, the diction is in a great measure casual and mutable; many of their terms are formed for some temporary or local convenience, and though current at certain times and places, are in others utterly unknown. This fugitive cant, which is always in a state of increase or decay, cannot be regarded as any part of the durable materials of a language, and therefore must be suffered to perish with other things unworthy of preservation.*" There's all sorts of dark matter in the English language.

See Samuel Johnson, A *Dictionary of the English Language* (London, 1755); *Merriam-Webster's Collegiate Dictionary*, 11th ed. (Springfield, MA: Merriam-Webster, 2003). We also recommend Pedro Carolino, *English As She Is Spoke* (New York: Appleton, 1883).

75 **Dark matter estimate.** We took a sample of a thousand words from a lexicon and determined how many fell into excluded categories. As a consequence, we don't have a list of all the English dark matter. Like the dark matter in the universe, we don't know exactly what it is—just that there's a lot of it.

Four Birthdays and a Funeral

76 **Word of the Year.** See "All of the Words of the Year, 1990 to Present," American Dialect Society, http://goo.gl/JCYMiK.

76 **Least Likely to Succeed.** We were thrilled to have beaten *skyaking*—jumping off a plane in a kayak—for this honor. It does seem to us, though, that given the mortal peril routinely faced by skyaking devotees, there might be a strong evolutionary argument that *skyaking* is indeed less likely to succeed. Of course, the ADS predictions should not be taken at face value; by 2011, *culturomics* had entered both the Random House and Macmillan dictionaries. See "Culturomics," *Macmillan Dictionary* online, http://goo.gl/qkg8GE; "Culturomics," Dictionary.com, http://goo.gl/EmvAhE.

77 **Chart.** Estimates for intermediate time points were based on linear interpolation.

78 **Causes of language growth and change.** It's fun to speculate about the exact causes of language change, and about the future of the English language in particular. See Michael Erard, "English As She Will Be Spoke," *New Scientist*, March 29, 2008; "English Is Coming," *Economist*, February 12, 2009, online at http://goo.gl/wcPGt8. People have been interested in this sort of thing for a long time. See Joseph Jacobs, "Growth of English—Amazing Development of the Language as Shown in the New Standard Dictionary's 450,000 Words," *New York Times*, November 16, 1913.

Daddy, where do babysitters come from?

79 **Two-to-one.** There are many examples of this sort of transition from two words to a compound word by means of a hyphenated intermediate. We don't want to railroad you with too many examples. See, for instance, NV: "rail road, rail-road, railroad."

CHAPTER 4. 7.5 MINUTES OF FAME

Cut the Crap

82 **Vatican Secret Archive.** The word *Secret—Segreto*—refers to the fact that the Archivio Segreto Vaticano is regarded as the personal property of the pope. That's not to say the place

isn't packed with juicy stuff, like a note from the English Parliament requesting a divorce for Henry VIII, the Papal Order excommunicating Martin Luther, and a letter announcing the abdication of the "hermaphrodite" Queen Christina of Sweden. Fortunately, in recent years, a massive cataloging effort has made its books a lot easier to find.

83 **Metadata quality.** An interesting, but now dated, thread on the early troubles Google encountered with book metadata can be found on the highly informative blog *Language Log*. See Geoff Nunberg, "Google Books: A Metadata Train Wreck," *Language Log*, August 29, 2009, http://goo.gl/AwNArh. The metadata quality has been improved dramatically since that time.

84 **Algorithms for improving metadata quality.** See Michel2011S.

Mr. Clean

85 **Ngrams vs. the human genome.** Estimates of genome base-call quality are based on Eric Lander et al., "Initial Sequencing and Analysis of the Human Genome," *Nature* 409, no. 6822 (2001): 860–921, online at http://goo.gl/trMZ4e.

85 **Ngrams and the law.** One emerging legal argument is that, whereas providing digital copies of millions of copyrighted texts for people to read ("consumptive" use) is a breach of copyright, making it possible to see the output of computations performed using those same copyrighted texts ("nonconsumptive" uses) may not be, so long as the output doesn't include long chunks of the original text. Ngrams are an example of a useful "nonconsumptive" use of books, a point we made in an amicus brief to the court in the case of *The Authors Guild, Inc., et al.*, v. *Google, Inc.* See Letter from Erez Lieberman-Aiden and Jean-Baptiste Michel to Court, September 3, 2009 (ECF No. 303), *The Authors Guild, Inc., et al.*, v. *Google, Inc.*, 770 F.Supp.2d 666 (S.D.N.Y., March 22, 2011) (No. 05-Civ.-8136).

This argument has gained some legal traction recently in the case of *The Authors Guild, Inc., et al.* v. *HathiTrust et al.* (S.D.N.Y., 2012). The HathiTrust Digital Library offers direct access to millions of digitized books obtained from participating libraries. Often these have been digitized by Google. On October 10, 2012, Hon. Harold Baer, Jr., a federal district judge in the Southern District of New York, ruled in favor of HathiTrust. The ruling specifically recognized that "nonconsumptive" computations over large collections of books constitute an "invaluable contribution to the progress of science and the cultivation of the arts" and that such benefits "fall safely within the protection of fair use." To support this view, Judge Baer cited an amicus brief filed by Matthew L. Jockers, Matthew Sag, and Jason Schultz, on which we were also signatories; as a specific example, the judge referred to the same ngram we used to open this book: "the frequency with which authors used 'is' to refer to the United States rather than 'are' over time." The ruling is online at http://goo.gl/QESiv; the amicus brief it cites is "Brief of Digital Humanities and Law Scholars as Amici Curiae in Partial Support of Defendants' Motion for Summary Judgment," *The Authors Guild, Inc., et al.*, v. *HathiTrust et al.*, 902 F.Supp.2d 445 (S.D.N.Y., October 10, 2012) (No. 11-Civ.-06351) 2012 WL 4808939.

What Fame Buys You

86 **Steven Pinker.** See *The Colbert Report*, 6:38, February 7, 2007, http://goo.gl/iFMGCt. Pinker was a coauthor on Michel2011.

The Story of Fame

88 **Most Googled person on Earth.** See "Zeitgeist 2010: How the World Searched," Google Zeitgeist, 2011, http://goo.gl/OCpY2X.

The Wright Stuff

89 **"I know it when I see it."** You'll know it when you see *Jacobellis* v. *Ohio*, 378 U.S. 184 (1963).

91 **Wind tunnels.** See Wilbur Wright et al., *The Papers of Wilbur and Orville Wright* (New

York: McGraw-Hill, 2000); Peter L. Jakab, *Visions of a Flying Machine: The Wright Brothers and the Process of Invention* (Washington, DC: Smithsonian Institution Press, 1990); Gina Hagler, *Modeling Ships and Space Craft: The Science and Art of Mastering the Oceans and Sky* (New York: Springer, 2013).

Almost Famous

95 **Michael Steele.** A video of the incident in question appears at "Steele Flubs 'Favorite Book' Reference During Debate," Newsmax, January 3, 2011, http://goo.gl/8hh40.

96 **Carol Gilligan.** See Andra Medea, "Carol Gilligan," *Jewish Women: A Comprehensive Historical Encyclopaedia*, http://goo.gl/LN2al.

Treating Fame Like a Disease

99 **Cohort method.** A translation of Andvord's 1930 original research appears at Kristian F. Andvord, "What Can We Learn by Following the Development of Tuberculosis from One Generation to Another?" *International Journal of Tuberculosis and Lung Disease* 6, no. 7 (2002): 562–68. For a survey of classic cohort studies, see Richard Doll, "Cohort Studies: History of the Method," *Sozial- und Präventivmedizin* 46, no. 2 (2001), 75–86, online at http://goo.gl/dRJKCp. The analyses in this chapter are all based on Michel2011 and are detailed there and in Michel2011S.

The Hall of Fame

101 **21758 Adrianveres.** Adrian's homeworld has an orbital period of 3.47 Earth years.

102 **7,500 victims.** Compiling a list of the fifty most famous people born each year between 1800 and 1950 involved a series of significant technical hurdles. One major problem was deciding when mentions of an ngram corresponding to the name of a person were actually referring to that person. Does the ngram *Winston Churchill* most likely refer to the statesman born in 1874, to his grandson born in 1940, to a novelist also named Winston Churchill and born in 1971, or to an impossible-to-disentangle mix of the three? To solve this problem, Veres used a great deal of contextual information, such as comparing the birthday of each Winston Churchill candidate with the ngram debut, noting the fact that the Wikipedia page for "Winston Churchill" redirects by default to the page of Winston1874, and observing that Winston1874 gets much more Wikipedia traffic than the other Winston Churchill candidates. These criteria and others were applied to hundreds of thousands of names. Read all about it at Michel2011S.

102 **Exciting set of people.** Later, Veres and *Science* journalist John Bohannon used ngrams to assemble a Science Hall of Fame comprising the most frequently mentioned contemporary scientists. They calculated the fame of each scientist in milliDarwins. One milliDarwin is one one-thousandth of the fame of Darwin. The most famous scientist turns out to be Bertrand Russell, whose antiwar positions made him the subject of great controversy. The most famous living scientist is Noam Chomsky, at 507 milliDarwins. See Adrian Veres and John Bohannon, "The Science Hall of Fame," *Science* 331, no. 6014 (January 14, 2011), online at http://goo.gl/6g8b7X.

The Grandee Unified Theory

108 **The dynamics of fame.** See Michel2011, Michel2011S.

How to Get Famous: A Guide to Choosing Your Career

113 **Celebrity focus group.** The list of the twenty-five most famous people born between 1800 and 1920 in each of the career categories can be consulted in its entirety at Michel2011S. The list features Marie Curie (1867, scientist), Marcel Duchamp (1887, artist), Claude Shannon (1916, mathematician), Humphrey Bogart (1899, actor), Virginia Woolf (1882, author), and Winston Churchill (1874, politician).

113 **On fame.** The study of fame is a well-established field of sociology. See Leo Braudy, *The Frenzy of Renown: Fame and Its History* (Oxford: Oxford University Press, 1986); Stanley Lieberson, *A Matter of Taste: How Names, Fashions, and Culture Change* (New Haven, CT: Yale University Press, 2000).

Infamy

118 **Chapman's words.** See Mark Sage, "Chapman Shot Lennon to 'Steal His Fame,'" *Irish Examiner,* October 19, 2004, online at http://goo.gl/pLXl51. A related controversy recently arose after *Rolling Stone* put a portrait of one of the Boston Marathon bombers, Dzhokhar Tsarnaev, on its cover. See Janet Reitman, "Jahar's World," *Rolling Stone,* July 17, 2013, http://goo.gl/fyc8y.

One giant leapfrog for mankind

120 **American heroes.** Raise your hand if you knew that the third astronaut on the mission—who orbited the moon in the command module while Armstrong and Aldrin were on the surface—was named Michael Collins.

CHAPTER 5. THE SOUND OF SILENCE

Intro

122 **"*Dort wo man Bücher verbrennt.*"** See Heinrich Heine, *Almansor,* in *Heinrich Heine's Gesammelte Werke,* ed. Carl Adolf Buchheim (Berlin: G. Grote, 1887); translation adapted from Stephen J. Whitfield, "Where They Burn Books," *Modern Judaism* 22, no. 3 (2002): 213–33, online at http://goo.gl/YbmMU3. Today, this passage appears in a memorial designed by Micha Ullman at Bebelplatz, a public square in Berlin, on the site where, during the 1933 book burnings, Joseph Goebbels led a mob in burning more than twenty thousand books. The memorial is a translucent pane in the square, through which onlookers can see enough empty bookshelves to accommodate twenty thousand books. You can see the inscription at http://goo.gl/SYzu4. Note that the passage from *Almansor,* as it appears in the Bebelplatz inscription, contains a typographic error.

123 **Helen Keller's letter.** A transcript of the letter, with changes written in by one of Keller's aides, gives insight into the editing process that led to the final version. It is in the collections of the American Foundation for the Blind, and can be seen at Helen Selsdon, "Helen Keller's Words: 80 Years Later . . . Still as Powerful," American Foundation for the Blind, May 9, 2013, http://goo.gl/uSSE8.

The annotations are discussed at Rebecca Onion, "'God Sleepeth Not': Helen Keller's Blistering Letter to Book-Burning German Students," *Slate,* May 16, 2013, http://goo.gl/SxdG2.

124 **Censorship.** See V. Gregorian, ed., *Censorship: 500 Years of Conflict* (New York: New York Public Library, 1984).

A Stained-Glass Window

124 **"Go and find a book."** See Jacob Baal-Teshuva, *Chagall: 1887–1985* (Cologne, Germany: Taschen, 2003), 16.

124 **Móyshe Shagal.** Although the name he ultimately adopted, Marc Chagall, was well established by 1910, he was known by many other names early on: Movsha Khatselev, Mark Zakharovich, Movsha Shagalov. See Benjamin Harshav, *Marc Chagall and His Times: A Documentary Narrative* (Palo Alto, CA: Stanford University Press, 2004), 63. Useful volumes about his life and art include Baal-Teshuva, above; Jackie Wullschlager, *Chagall: A Biography* (New York: Alfred A. Knopf, 2008); Marc Chagall, *The Jerusalem Windows,* trans. Jean Leymarie (New York: George Braziller, 1967); Marc Chagall, *My Life,* trans. Elisabeth Abbott (New York: Da Capo Press, 1994).

125 **"The quintessential Jewish artist."** See Robert Hughes, "Fiddler on the Roof of Modernism," *Time*, June 24, 2001, http://goo.gl/aFMsU.

125 **"When Matisse dies."** See Françoise Gilot and Carlton Lake, *Life with Picasso* (New York: McGraw-Hill, 1964), 258. Gilot was Picasso's lover and muse. She notes that, although Picasso had some personal issues with Chagall, he nonetheless had immense respect for Chagall's art. The full quotation is: "*When Matisse dies, Chagall will be the only painter left who understands what color really is. I'm not crazy about those cocks and asses and flying violinists and all the folklore, but his canvases are really painted, not just thrown together. Some of the last things he's done in Venice convince me that there's never been anybody since Renoir who has the feeling for light that Chagall has.*"

125 **Commissar for the visual arts.** See Wullschlager, 223.

125 **"I'm afraid that my 'image.'"** See Harshav, 326–27.

126 **Chart.** NV: "Chagall"/French, "Шагал"/Russian.

Degenerate Art

127 **Max Nordau.** His views on degenerate art appear in the two-volume *Entartung* [*Degeneration*] (Berlin: Carl Dunder Verlag, 1892–1893). The Nazi usage of this concept was obviously a 180-degree reversal of Nordau's broader views. See, for instance, Max Nordau and Gustav Gottheil, *Zionism and Anti-Semitism* (New York: Fox, Duffield, 1905); Max Nordau and Anna Nordau, *Max Nordau: A Biography* (Whitefish, MT: Kessinger, 2007). Nordau was vice president of the first six World Zionist Congresses (Theodor Herzl was president), and president of the next four. See "Max Nordau," *The Encyclopedia of the Arab-Israeli Conflict*, ed. Spencer C. Tucker (Santa Barbara, CA: ABC-CLIO, 2008).

127 **Draconian control of German culture.** See Richard A. Etlin, *Art, Culture, and Media Under the Third Reich* (Chicago: University of Chicago Press, 2002); Glenn R. Cuomo, ed., *National Socialist Cultural Policy* (New York: St. Martin's Press, 1995); Alan E. Steinweis, *Art, Ideology, and Economics in Nazi Germany* (Chapel Hill: University of North Carolina Press, 1993); Jonathan Petropoulos, *The Faustian Bargain* (New York: Oxford University Press, 2000).

128 **"In the future, only those."** Peter Adam, *Art of the Third Reich* (New York: Harry N. Abrams, 1992), 53.

128 ***The Scream.*** The museum did not agree to do so. See Marcy Oster, "Heirs of Owner of Nazi-Looted 'The Scream' Want Explanation on Display at MoMA," Jewish Telegraphic Agency, October 15, 2012, http://goo.gl/gBmtL.

The Most Popular Art Exhibit of All Time

129 **"German *Volk*."** The translation is drawn from Neil Levi, "'Judge for Yourselves!'—The 'Degenerate Art' Exhibition as Political Spectacle," *October* 85 (1998): 41–64, online at http://goo.gl/CfuBMt.

129 ***Entartete Kunst.*** In 1991, Stephanie Barron curated a reconstruction of *Entartete Kunst* for an exhibition at the Los Angeles County Museum of Art. The catalog she created for this exhibition is an invaluable scholarly contribution. See Stephanie Barron, ed., *Degenerate Art: The Fate of the Avant-garde in Nazi Germany* (Los Angeles: Los Angeles County Museum of Art, 1991).

131 **"I felt an overwhelming sense of claustrophobia."** The quote is from "Three Days in Munich, July 1937," an essay by Peter Guenther that appears in Barron's catalog. This fascinating document describes Guenther's visits to the *Große Deutsche Kunstausstellung* and *Entartete Kunst* as a seventeen-year-old. See ibid., 38.

131 **Most popular art exhibition.** On August 2, 1937, alone, thirty-six thousand people attended *Entartete Kunst*. To give a sense of how massive this turnout was, it's useful to examine worldwide exhibition attendance statistics that are conveniently available from the *Art Newspaper* (www.theartnewspaper.com) for the past ten years. The statistics for 20XX are available at http://www.theartnewspaper.com/attfig/attfigXX.pdf. Notably, only one of the

exhibits listed exceeded the daily average attendance of *Entartete Kunst* over the latter's first four months. The exception was a 2009 exhibition of the Shoso-In treasures of Emperor Shōmu (701–756) and Empress Kōmyō (701–760) in Nara, Japan, which sustained an average daily attendance of 17,926 people. However, the exhibit was only up for about two weeks, and thus total attendance, at a little more than a quarter million people, was a small fraction of the attendance of *Entartete Kunst*. In general, there are some shows with very high attendance over an extremely brief period of time, but none that comes close to matching the sustained interest achieved by *Entartete Kunst*. The claim that "the popularity of *Entartete Kunst* has never been matched by any other exhibition of modern art" is explicitly made in Barron, 9; although we obviously do not have attendance figures for every art exhibition in history, this claim seems very plausible to us based on available figures.

132 **Emil Nolde.** Nolde was a supporter of the Nazi Party, but was nevertheless a target because of Hitler's rejection of Expressionism.

Book Burnings

133 **Posters.** The poster can be seen at http://goo.gl/bNK9H.

133 **"We want to regard."** This translation is from "List of Banned Books, 1932–1939," University of Arizona, June 22, 2002, http://goo.gl/PMVRy.

134 **Blacklists.** The blacklists are detailed in W. Treß, *Wider den Undeutschen Geist: Bücherverbrennung 1933* (Berlin: Parthas, 2003); G. Sauder, *Die Bücherverbrennung: 10. Mai 1933* (Frankfurt am Main: Ullstein, 1985); and *Liste des Schädlichen und Unerwünschten Schrifttums* (Leipzig: Hedrich, 1938).

Communications with W. Treß and the City of Berlin Web site (berlin.de) provided us with immense assistance in creating digital versions of the blacklists. A very helpful timeline appears at http://goo.gl/0ig7Ig.

135 **The work of Margaret Stieg Dalton.** See Margaret F. Stieg, *Public Libraries in Nazi Germany* (Tuscaloosa: University of Alabama Press, 1992) and Alan E. Steinweis, review of *Public Libraries in Nazi Germany*, by Margaret F. Stieg, DigitalCommons@University of Nebraska-Lincoln, April 1, 1992, http://goo.gl/atlK2t.

What They Don't Want You to Know: A World Tour

137 **Suppression in Russia.** See Robert Service, *Stalin: A Biography* (Cambridge, MA: Harvard University Press, 2004). Stalin didn't just manage to edit rivals out of the textual record. He was also, for instance, very aggressive about having his rivals doctored out of photographs. See David King, *The Commissar Vanishes* (New York: Metropolitan Books, 1997); Joseph Gibbs, *Gorbachev's Glasnost* (College Station: Texas A&M University Press, 1999).

138, 142 **Charts.**

NV: "**Троцкий, Зиновьев, Каменев**"/Russian (smoothing = 1).

NV: "Tiananmen"/English, "天安门"/Chinese (smoothing = 0). Note that the axes are on different scales; the exact query is: "Tiananmen:eng_2012 * 10, 天安门:chi_sim_2012." Spurious peaks prior to 1950 are due to the small number of books written before that date in the Chinese corpus. Chinese sources tend to refer to these events as "the June 4th Incident," 六四事件. Indeed, NV: "六四事件"/Chinese shows a rise at the expected time; however, this is not surprising, given that this phrase has no referent before 1989.

139 **The Hollywood Ten.** For portraits of the Hollywood Ten, see Bernard F. Dick, *Radical Innocence* (Lexington: University Press of Kentucky, 1988); Gerald Horne, *The Final Victim of the Blacklist* (Berkeley: University of California Press, 2006); the autobiographical Edward Dmytryk, *Odd Man Out* (Carbondale: Southern Illinois University Press, 1996); and the remarkable documentary film *The Hollywood Ten*, directed by John Berry, 1950.

140 **"Until such time as he is acquitted."** The full text of the Waldorf Statement appears

in William T. Walker, *McCarthyism and the Red Scare* (Santa Barbara, CA: ABC-CLIO, 2011), 136.

140 **"Most un-American thing in the country today."** See Jonathan Auerbach, *Dark Borders* (Durham, NC: Duke University Press, 2011), 4.

141 ***Exodus.*** See *Exodus*, directed by Otto Preminger, 1960.

141 **Tiananmen Square.** For more about the massacre, see Dingxin Zhao, *The Power of Tiananmen* (Chicago: University of Chicago Press, 2001); Scott Simmie and Bob Nixon, *Tiananmen Square* (Seattle: University of Washington Press, 1990); Philip J. Cunningham, *Tiananmen Moon* (Lanham, MD: Rowman & Littlefield, 2009); Timothy Brook, *Quelling the People* (Palo Alto, CA: Stanford University Press, 1992).

142 **"Great Firewall of China."** See Xiao Qiang and Sophie Beach, "The Great Firewall of China," *St. Petersburg Times*, September 3, 2002; "The Great Firewall: The Art of Concealment," *Economist*, April 6, 2013, http://goo.gl/VTV3b.

The Chinese effort to censor search engines such as Google brings us back, in some ways, to the notion of a concordance or card catalog. If you can't get rid of the contents of the library (in this analogy, by shutting down the entire Internet), you can effectively restrict access by eliminating the concordance or the card catalog (search engines that help you find the page or word you're interested in.) For more about censorship of and by Google in China, see "Google Censors Itself for China," BBC, January 25, 2006, http://goo.gl/Xydlua; Michael Wines, "Google to Alert Users to Chinese Censorship," *New York Times*, June 1, 2012, http://goo.gl/7QmrQ; Josh Halliday, "Google's Dropped Anti-Censorship Warning Marks Quiet Defeat in China," *Guardian*, January 7, 2013, http://goo.gl/aA2HU.

For more about Chinese Internet censorship of the Tiananmen Square massacre, see Jonathan Kaiman, "Tiananmen Square Online Searches Censored by Chinese Authorities," *Guardian*, June 4, 2013, http://goo.gl/60SIo; Matt Schiavenza, "How China Made the Tiananmen Square Massacre Irrelevant," *Atlantic*, June 4, 2013, http://goo.gl/d7Ccw. For more about the Tank Man, see Patrick Witty, "Behind the Scenes: Tank Man of Tiananmen," *New York Times*, June 3, 2009, http://goo.gl/IvhdX.

Perhaps the most telling insights come from asking members of younger generations in China what they know about the incident, when they learned about it, and how they found out, as in "China's Tiananmen Generation Speaks," BBC, May 28, 2009, http://goo.gl/ms7x2, and in "Chinese Students Unaware of the 'Tank Man,'" *Frontline*, video, 2:37, July 27, 2008, http://goo.gl/Jf0Hy.

Can We Detect Censorship Automatically?

143 **Detecting censorship.** See Michel2011 and Michel2011S for details.

Seeping Through a Million Channels

147 **Mocking can be marketing.** When the Nazis followed up the *Entartete Kunst* exhibit with concerts featuring jazz, Jewish songs, and other *entartete Musik*, they became increasingly worried about this form of subversion, suspecting that the listeners in attendance were actually coming because they were fans of the music. See Michael Haas, *Forbidden Music* (New Haven, CT: Yale University Press, 2013); "Music in the Third Reich," Music and the Holocaust, http://goo.gl/OlNcwZ.

147 **Charlotte Salomon.** See Charlotte Salomon, *Life? or Theater?*, trans. Leila Vennewitz (New York: Viking, 1981); Mary Lowiner Felstiner, *To Paint Her Life* (New York: Harper Perennial, 1995); Michael P. Steinberg and Monica Bohm-Duchen, *Reading Charlotte Salomon* (Ithaca, NY: Cornell University Press, 2006).

148 **"The pictorial counterpart of Anne Frank's diary."** See "A Poignant Reminder of the Value of Life," *St. Petersburg Times*, October 6, 1963.

149 **"So tenderly."** The quote is from Paula Salomon-Lindberg, Charlotte's stepmother. See Felstiner, 228.

Two rights make another right

150 This ngram was originally pointed out by Steven Pinker, and is discussed in greater detail in Steven Pinker, *The Better Angels of Our Nature: Why Violence Has Declined* (New York: Viking, 2011).

CHAPTER 6. THE PERSISTENCE OF MEMORY

Intro

152 **Vienna Circle.** See Thomas Uebel, "Vienna Circle," *The Stanford Encyclopedia of Philosophy* (Summer 2012); Alfred J. Ayer, *Logical Positivism* (Glencoe, IL: Free Press, 1959); Friedrich Weismann et al., *Wittgenstein and the Vienna Circle* (Oxford: Basil Blackwell, 1979); and David Edmonds and John Eidinow, *Wittgenstein's Poker* (New York: Ecco, 2001).

153 **Opposition to the term *Volksgeist*.** See Verein Ernst Mach, *Wissenschaftliche Weltauffassung: Der Wiener Kreis* (Vienna: Artur Wolf, 1929).

Memory Test

154 **Ebbinghaus.** See Hermann Ebbinghaus, *Memory: A Contribution to Experimental Psychology*, trans. Henry Ruger and Clara Bussenius (1885; New York: Teachers College, Columbia University, 1913). William James' glowing review of this work can be found in William James, *Essays, Comments and Reviews* (Cambridge, MA: Harvard University Press, 1987). Although Ebbinghaus was a pioneer of experimental psychology, he was not among the very first wave; significant figures predating Ebbinghaus include Wilhelm Wundt, often regarded as the father of experimental psychology, and William James, mentioned above, often regarded as the father of American psychology.

Unforgettable

158 **Chart.** NV: "Lusitania, Pearl Harbor, Watergate" (smoothing = 0).

A Memory by Any Other Name

160 **1876.** The probability that a given digit or that a given number appears in a text is not uniform. Instead, it follows a heavy-tailed distribution—similar in some respects to a power law—called Benford's law. See, for instance, Theodore P. Hill, "A Statistical Derivation of the Significant Digit Law," *Statistical Science* 10, no. 4 (November 1995): 354–63, online at http://goo.gl/hLtUvm.

According to Benford's law, the likelihood of seeing the number 1876 in a text is virtually nil. In fact, we see this and nearby numbers in very significant quantities, an otherwise anomalous finding that makes perfect sense in light of the fact that they predominantly correspond to years.

Benford's law is a remarkably powerful observation. For instance, it can be used to detect fraud in tax returns: When fabricating numbers, people tend not to follow Benford's law. This application was suggested by, among others, Hal Varian, currently chief economist at Google. See Hal Varian, "Letters to the Editor," *American Statistician* 26, no. 3 (June 1972). For more on the relationship between the mind and numbers, see Stanislas Dehaene, *The Number Sense: How the Mind Creates Mathematics* (Oxford: Oxford University Press, 1997).

Out with the Old, In with the New

164 **The speed of information before the information age.** William Dockwra established the

Penny Post in London in 1680, advertising "For One Penny" delivery "at least fifteen times per day" to "places of quick negotiation within the City," starting at 6:00 a.m. and ending at 9:00 p.m., or about once every hour. He also promised delivery at least five times a day "to the most remote places" in and around London, and the Penny Post guaranteed delivery in four hours or less. It would be great if the post office could do that today. Read the advertisement yourself at "London Penny Post," The British Postal Museum & Archive, http://goo.gl/qwAtI. See Catherine Golden, *Posting It: The Victorian Revolution in Letter Writing* (Gainesville: University Press of Florida, 2009); George Brumell, *The Local Posts of London 1680–1840* (Cheltenham, England: R. C. Alcock, 1950); "Provincial Penny Post/5th Clause," The British Postal Museum & Archive, http://goo.gl/jomYJ; Randall Stross, "The Birth of Cheap Communication (and Junk Mail)," *New York Times*, February 20, 2010, online at http://goo.gl/SO0L0Y; Robert Darnton, "An Early Information Society: News and the Media in Eighteenth-Century Paris," *American Historical Review* 105, no. 1 (February 2000).

Buckminster Fuller created a beautiful graphical representation of the maximal speed at which information could travel throughout history. See Buckminster R. Fuller and John McHale, "Shrinking of Our Planet," online at http://goo.gl/IfvqBL.

It's not just pure information that moved quickly in earlier times. In the nineteenth century, physical parcels could be mailed from place to place in a city through—quite literally—a network of underground tubes. These pneumatic tubes used air pressure to deliver parcels all over cities like New York and Paris at speeds of up to twenty-five miles per hour. They were organized into vast, complex tube networks that made their way through large parts of many major cities. New York discontinued use of pneumatic mail in the '50s. Paris kept its system working through the '80s, when it was largely replaced by the use of fax machines. Today, we do indeed live in an information age, in which we've become phenomenally good at moving around information. But if you wanted to send an actual pineapple across Manhattan, instead of just sending a gif of a pineapple or a letter about a pineapple, it's quite possible that you would have been better off living a century ago.

Presumably, these tubes still exist, and we imagine that rodents must inhabit them from time to time. Thus it is fair to say that underneath New York lies an information superhighway consisting of squirrels running around in tubes. It's just not the Internet. (And it's probably rats, not squirrels.) See: J. D. Hayhurst, *The Pneumatic Post of Paris* (Oxford: France and Colonies Philatelic Society of Great Britain, 1974); L. C. Stanway, *Mails Under London: The Story of the Carriage of the Mails on London's Underground Railways* (Basildon, England: Association of Essex Philatelic Societies, 2000); "Pneumatic Mail," National Postal Museum, http://goo.gl/uwsgmz.

Notably, Elon Musk, the entrepreneur behind PayPal, Tesla Motors, and SpaceX, recently proposed bringing back pneumatic tube transport for both humans and cargo, an approach to mass transit that he has dubbed the Hyperloop. See Damon Lavrinc, "Elon Musk Thinks He Can Get You from NY to LA in 45 Minutes," CNN Tech, July 17, 2013, http://goo.gl/EXPdT.

Eureka

166 **The fax machine.** How is it possible that the fax was invented before the telephone? Adequately encoding the richness of a human voice was likely a greater challenge than encoding geometric figures.

Patent Caveat

167 **Who invented the phone?** The debate over who deserves the title of "inventor of the telephone" is still raging. In 2002, the United States House of Representatives voted to recog-

nize Antonio Meucci as an inventor of the telephone. Meanwhile, the Canadian government officially declared that the evidence was not substantial enough to support Meucci's claim. We hope the United Nations Security Council will weigh in soon. See Robert V. Bruce, *Bell: Alexander Graham Bell and the Conquest of Solitude* (Boston: Little, Brown, 1973). For more on Meucci, see *Scientific American Supplement*, no. 520 (December 19, 1885).

168 **"I then shouted into M."** See the Alexander Graham Bell Family Papers at the Library of Congress, 1862–1939, http://memory.loc.gov/ammem/bellhtml/.

147 Blind Dates

170 **List of inventions.** The full list of inventions we used for this study is available in Michel2011S. There is invariably a lag, usually several years long, between when something was invented and when the patent was issued. In some cases, the date of invention could be unambiguously determined, and there was a particularly long delay before the patent was granted. An example is the theremin, a musical instrument invented in 1920 by Leon Theremin in Russia; the U.S. patent for the device was issued in 1928. In such cases, we use the date of invention, not the date on which the patent was issued.

171 **The life cycle of inventions.** Everett M. Rogers, *Diffusion of Innovations* (New York: Free Press, 1962) is a classic text on the way innovation spreads through a society.

Singularity or Bust!

173 **Ulam and von Neumann.** The quote is from a moving obituary of von Neumann written by Stanislaw Ulam, in which Ulam recalls a discussion with von Neumann on this topic. The obituary provides a broad review of von Neumann's numerous visionary contributions to modern science. See Stanislaw Ulam, "John von Neumann 1903–1957," *Bulletin of the American Mathematical Society* 64 (1958): 1–49.

174 **Ray Kurzweil.** See his book *The Singularity Is Near: When Humans Transcend Biology* (New York: Viking, 2005). Since 2012, Kurzweil has been the director of engineering at Google, with a mandate to make computers understand natural language.

Volksgeist, *Culture, Culturomics*

175 **Johann Gottfried Herder.** In addition to the term *Volksgeist*, Herder also coined the widely used term *Zeitgeist*, or "spirit of the time." See Johann Gottfried Herder, *Reflections on the Philosophy of the History of Mankind* (Chicago: University of Chicago Press, 1968); Frederick M. Barnard, *Herder's Social and Political Thought* (Oxford: Clarendon Press, 1965).

175 **Herder, nationalism, and racism.** See Robert Reinhold Ergang, *Herder and the Foundations of German Nationalism* (New York: Columbia University Press, 1931); George M. Fredrickson, *Racism: A Short History* (Princeton, NJ: Princeton University Press, 2003); Eve Garrard and Geoffrey Scarrey, eds., *Moral Philosophy and the Holocaust* (Burlington, VT: Ashgate, 2003).

176 **Franz Boas.** Of course, Boas' take on culture was bad business for hate-mongers. The Nazis burned his books, rescinded his PhD, and denounced Boasian anthropology as "Jewish science." For more on Boas' contributions to the concept of culture, see George W. Stocking, Jr., "Franz Boas and the Culture Concept in Historical Perspective," *American Anthropologist* 68 (1966): 867–82, online at http://goo.gl/VIyZ8g. Also see Stocking's edited volume *Volksgeist as Method and Ethic: Essays on Boasian Ethnography and the German Anthropological Tradition* (Madison: University of Wisconsin Press, 1998). In particular, see Matti Bunzl's contribution to that book, "Franz Boas and the Humboldtian Tradition: From *Volksgeist* and *Nationalcharakter* to an Anthropological Notion of Culture."

176 **-omics.** When we coined the term "culturomics," we always intended for it to be pronounced with a long *o*, as in the standard pronunciation of *genomics* (or as in the word

"owe"). Recently, however, the pronunciation guide of the Macmillan dictionary reported that the word ought to be pronounced with a short *o*, as in *economics*. (See notes to "Four Birthdays and a Funeral," above.) Can the dictionary be wrong about something like this? Did we get it wrong? Were we mispronouncing it from the beginning, or did we start being wrong only after Macmillan made its announcement? For more about *-omics*, see James Gorman, "'Ome,' the Sound of the Scientific Universe Expanding," *New York Times*, May 3, 2012, http://goo.gl/I0um5.

Coping with Addiction: A New Strategy

180 **Ngram Viewer.** We would like to apologize to everyone for creating such an effective time-waster. It was never our intention to waste so much of people's time. If only there were some way we could undo the damage caused by all that lost productivity. For more on how the Ngram Viewer has been used, see Patricia Cohen, "In 500 Billion Words, a New Window on Culture," *New York Times*, December 16, 2010, online at http://goo.gl/16gtxR; Alexis C. Madrigal, "Vampire vs. Zombie: Comparing Word Usage Through Time," *Atlantic*, December 17, 2010, online at http://goo.gl/MUUnG1.

Mommy, where do Martians come from?

182 **Galileo.** Galileo discusses this point in *Dialogue Concerning the Two Chief World Systems*, 321. For a modern attempt to reconstruct some of Galileo's Martian observations, see William T. Peters, "The Appearances of Venus and Mars in 1610," *Journal for the History of Astronomy* 15, no. 3 (1984).

183 **Schiaparelli.** See Giovanni Virginio Schiaparelli, *La Vita sul Pianeta Marte* (Milan: Associazione Culturale Mimesis, 1998).

183 **Canals on Mars.** Lowell's original three books on the topic are *Mars* (Boston: Houghton Mifflin, 1895); *Mars and Its Canals* (New York: Macmillan, 1911); and *Mars as the Abode of Life* (New York: Macmillan, 1908). Alfred Russel Wallace, in *Is Mars Habitable?* (New York: Macmillan, 1907), refuted Lowell's position. See also Steven J. Dick, *Life on Other Worlds* (Cambridge: Cambridge University Press, 1998); Robert Markley, *Dying Planet* (Durham, NC: Duke University Press, 2005). For more about Lowell, see David Strauss, *Percival Lowell* (Cambridge, MA: Harvard University Press, 2001).

183 **Dean of American astronomers.** See David H. Devorkin, *Henry Norris Russell: Dean of American Astronomers* (Princeton, NJ: Princeton University Press, 2000).

183 **"Perhaps the best."** See Dick, *Life on Other Worlds*, 35.

184 ***The War of the Worlds.*** See H. G. Wells, *The War of the Worlds* (London: William Heinemann, 1898).

184 **The Martian globe used to plan the Mariner missions.** The globe was based on a map known as the MEC-1 prototype, created by E. C. Slipher, who had trained under Lowell. Despite the scientific consensus having turned against canals, Slipher appears to have remained bullish about them until his death in 1964. The *Mariner 4* flyby took place in 1965. You can see the MEC-1 prototype map at http://goo.gl/GrOKZ, and you can even explore the maps of Martian canals using Google Earth. For a video that describes how, see "Mars," Google Earth, http://goo.gl/ZXZZa. Slipher's collected papers are at "E. C. Slipher Collection," Arizona Archives Online, http://goo.gl/jXva1D.

184 **Mariner.** For more on the Mariner missions, see John Hamilton, *The Mariner Missions to Mars* (Minneapolis: ABDO, 1998).

CHAPTER 7. UTOPIA, DYSTOPIA, AND DAT(A)TOPIA

Intro

185 **King David.** See II Samuel 24.

Digital Past

186 **Edgar Allan Poe.** See Jeffrey Meyers, *Edgar Allan Poe: His Life and Legacy* (New York: Charles Scribner's Sons, 1992). A low-resolution facsimile of Poe's balloon hoax appears at "Réseau Pneumatic de Paris," *Cix*, 2000, http://goo.gl/nCo3s.

187 **Update to the Ngram Viewer.** The most recent version of the ngram data draws from eight million books and introduces part-of-speech tagging. See Yuri Lin et al., "Syntactic Annotations for the Google Books Ngram Corpus," *Proceedings of the ACL 2012 System Demonstrations* (2012): 169–74; Yuri Lin, "Syntactically Annotated Ngrams for Google Books" (master's thesis, Massachusetts Institute of Technology, 2012).

187 **The number of books digitized by Google.** See Robert Darnton, "The National Digital Public Library Is Launched!," *New York Review of Books*, April 25, 2013, online at http://goo.gl/OI5n2J.

187 **E-books.** By 2009, Amazon was already selling more e-books than hardcover books. See Charlie Sorrel, "Amazon: Kindle Books Outsold Real Books This Christmas," *Wired*, December 28, 2009, online at http://goo.gl/ZsB7it. In 2012, e-books accounted for 23 percent of the book market in the United States. See Jeremy Greenfield, "Ebooks Account for 23% of Publisher Revenue in 2012, Even as Growth Levels," *Digital Book World*, April 11, 2013, online at http://goo.gl/u0d1GJ.

187 **Increasing access to digital books.** The HathiTrust (http://www.hathitrust.org), the Internet Archives (http://archive.org/index.php), Project Gutenberg (http://www.gutenberg.org), and the Digital Public Library of America (http://dp.la) are several of the most notable efforts aimed at making digital books available to the public. When the full texts of books are available, one can build far more powerful tools for cultural trends.

An example can be found at bookworm.culturomics.org. Google's closed-source adaptation of the original Bookworm uses the name "Ngram Viewer." "Bookworm" is an open-source effort at the Cultural Observatory. The Bookworm code base was developed together with Benjamin Schmidt, Neva Cherniavsky Durand, Martin Camacho, Matthew Nicklay, and Linfeng Yang. Schmidt was the lead developer.

189 **The threat to libraries.** See S. Peter Davis, "6 Reasons We're in Another 'Book-Burning' Period in History," *Cracked*, October 11, 2011, http://goo.gl/FBZoD; Matthew Shaer, "Dead Books Club," *New York*, August 12, 2012, http://goo.gl/UAIDN; Mari Jones, "David Lloyd George's Books Pulped by Conwy Libraries Services," *Daily Post*, March 24, 2011, http://goo.gl/b1pK0; Helen Carter, "Authors and Poets Call Halt to Book Pulping at Manchester Central Library," *Guardian*, June 22, 2012, http://goo.gl/lEas1P.

190 **Newspaper digitization.** See Chronicling America, National Endowment for the Humanities, http://chroniclingamerica.loc.gov; Trove, National Library of Australia, http://trove.nla.gov.au; and the now-defunct effort Google News Archive, Google News, http://news.google.com/newspapers.

191 **Ancient and unpublished texts.** See, for instance, "Digitized Dead Sea Scrolls," Israel Museum, Jerusalem, http://dss.collections.imj.org.il; Perseus Digital Library, Tufts University, http://www.perseus.tufts.edu. An effort to digitize artifacts related to Poe can be found at "The Edgar Allan Poe Digital Collection," Harry Ransom Center, University of Texas at Austin, http://goo.gl/XvcqO.

192 **Digitizing the physical world.** See Europeana, http://europeana.eu, for a vast effort at opening access to texts, artwork, films, and many other cultural objects in Europe.

Digital Present

194 **Our data footprint.** See Josh James, "How Much Data Is Created Every Minute?," *DOMO*, June 8, 2012, http://goo.gl/RN5eB. Professor Gregory Crane, editor in chief of the Perseus Library Project, aiming to digitize all texts from ancient Greece, suggested that roughly one hundred million words of Greek survive from before 600 CE; Gregory Crane, e-mail to Jean-Baptiste Michel, May 18, 2013.

194 **Spam.** Of the 107 trillion e-mails sent in 2010, 89.1 percent were spam. See "Internet 2010 in Numbers," *Royal Pingdom*, January 12, 2011, online at http://goo.gl/ziXncU.

Digital Future

197 **TotalRecall.** Deb Roy's TED talk is entertaining and informative. See Deb Roy, *The Birth of a Word*, video, 19:52, March 2011, http://goo.gl/5MoJo. More details about the project are available at Jonathan Keats, "The Power of Babble," *Wired*, March 2007, http://goo.gl/3epTR; Jason B. Jones, "Making That Home Video Count," *Wired*, March 25, 2011, http://goo.gl/V3oTL. More technical overviews include Deb Roy et al., "The Human Speechome Project," Massachusetts Institute of Technology, July 2006, http://goo.gl/O3E0e; Rony Kubat et al., "TotalRecall: Visualization and Semi-Automatic Annotation of Very Large Audio-Visual Corpora," Massachusetts Institute of Technology, http://goo.gl/Dra7T.

198 **Life logging.** Life logging, wearable computing, and the increasingly fashionable notion of a quantified self are all intimately related concepts. See Steve Henn, "Clever Hacks Give Google Many Unintended Powers," NPR, July 17, 2013, http://goo.gl/eyUW9; Edna Pasher and Michael Lawo, *Intelligent Clothing* (Lansdale, PA: IOS Press, 2009); Tomio Geron, "Scan Your Temple, Manage Your Health with New Futuristic Device," *Forbes*, November 29, 2012, http://goo.gl/9lg72; Greg Beato, "The Quantified Self," *Reason*, December 21, 2011; Mark Krynsky, "The Best Health and Fitness Gadget Announcements from CES 2013," Lifestream Blog, January 18, 2013, http://goo.gl/Qq0BY; Eric Topol, *The Creative Destruction of Medicine* (New York: Basic Books, 2011); Jody Ranck, *Connected Health* (San Francisco: GigaOM, 2012).

199 **Mind-machine interfaces.** See Leigh R. Hochberg et al., "Neuronal Ensemble Control of Prosthetic Devices by a Human with Tetraplegia," *Nature* 442, no. 7099 (2006): 164–71; Martin M. Monti et al., "Willful Modulation of Brain Activity in Disorders of Consciousness," *New England Journal of Medicine* 362, no. 7 (2010): 579–89. Both are landmark studies.

200 **Stream of consciousness.** See Steven Pinker, *The Stuff of Thought* (New York: Viking Penguin, 2007), and Chris Swoyer, "Relativism," *The Stanford Encyclopedia of Philosophy* (Winter 2010). The notion of a stream of consciousness is generally credited to William James.

Truth and Consequences

201 **The Boston Marathon bombing investigation.** Investigators combed through vast quantities of pictures and movies recorded by individuals present on the scene and asked the public for help identifying two suspects. See Spencer Ackerman, "Data for the Boston Marathon Investigation Will Be Crowdsourced," *Wired*, April 16, 2013, online at http://goo.gl/DpPKca; Pete Williams et al., "Investigator Pleads for Help in Marathon Bombing Probe: 'Someone Knows Who Did This,'" NBC News, April 16, 2013, online at http://goo.gl/46kndz.

202 **Rehtaeh Parsons.** The seventeen-year-old hanged herself on April 4, 2013. As a result, she fell into a coma; three days later, she was taken off life support. See "Rehtaeh Parsons, Canadian Girl, Dies After Suicide Attempt; Parents Allege She Was Raped by 4 Boys," *Huffington Post*, April 9, 2013, online at http://goo.gl/Cqs030.

Data Is Power

203 **What marketers know about you.** See Charles Duhigg, "How Companies Learn Your Secrets," *New York Times*, February 16, 2012, online at http://goo.gl/DV04Me.

204 **What the government knows about you.** See Joseph Ax, "Occupy Wall Street Protester Can't Keep Tweets from Prosecutors," *Chicago Tribune*, September 17, 2012.

205 **Offlogging.** See Jamie Skorheim, "Seattle Bar Steps Up as First to Ban Google Glasses," MyNorthwest.com, March 8, 2013.

205 **Snapchat.** Note that Snapchat's "deleted" messages can be recovered, at least in some cases; this discovery has led to a formal complaint to the Federal Trade Commission. See Jessica Guynn, "Privacy Watchdog EPIC Files Complaint Against Snapchat with FTC," *Los Angeles Times*, May 17, 2013, http://goo.gl/WSxTxA.

Kindred Spirits

206 **Pioneers.** See Franco Moretti, *Graphs, Maps, Trees: Abstract Models for a Literary History* (London: Verso, 2005), and, in this vein, the quote cited above (in the notes to "Taking Roses Apart to Count Their Petals") by George Miller; Matthew L. Jockers, *Macroanalysis: Digital Methods and Literary History* (Urbana: University of Illinois Press, 2013); James M. Hughes et al., "Quantitative Patterns of Stylistic Influence in the Evolution of Literature," *Proceedings of the National Academy of Sciences* 109, no. 20 (2012): 7682–86, online at http://goo.gl/3uaAoM; James W. Pennebaker, *The Secret Life of Pronouns: What Our Words Say About Us* (New York: Bloomsbury, 2011). The Shared Horizons conference Web site is at http://goo.gl/fnyWw. For an insightful read about the future of science and the humanities, we recommend Edward O. Wilson, *Consilience: The Unity of Knowledge* (New York: Alfred A. Knopf, 1998). The essential reference on the tension between the sciences and the humanities is C. P. Snow, *The Two Cultures and the Scientific Revolution* (London: Cambridge University Press, 1959).

Psychohistory

209 **The quantified society.** See Adolphe Quetelet, *Sur l'Homme et le Développement de Ses Facultés, ou, Essai de Physique Sociale* (Brussels: L. Hauman, 1836); Émile Durkheim, *Les Règles de la Méthode Sociologique* (Paris: F. Alcan, 1895); Auguste Comte and Harriet Martineau, *The Positive Philosophy* (New York: AMS Press, 1974). It's worth comparing this line of thought with that which motivated Zipf, 1935:

> *Nearly ten years ago, while studying linguistics at the University of Berlin, it occurred to me that it might be fruitful to investigate speech as a natural phenomenon . . . in the manner of the exact sciences, by the direct application of statistical principles to the objective speech-phenomena.*

211 **Chart.** We analyzed cultural inertia with the help of Martin Camacho and Guillaume Basse, both students at Harvard. We asked whether ngrams that increase linearly, leading to doubling over two decades, continue to rise after the initial two-decade period. Hundreds of such ngrams are averaged to create the dark gray line shown in the plot; each point in the plot is the median of all the ngrams included in the average at that time point. Note that the time axis for each ngram is offset so that the initial twenty-year rise always begins at Year 0. This initial twenty-year period, during which a sharp increase is guaranteed because of how the ngrams were selected, is highlighted. Subsequently, the ngrams continue to rise, indicating inertia. The ngrams averaged in light gray were selected based on a twenty-year-long linear decrease. These also show inertia, this time in the downward direction. The effect is very pronounced. Although it cannot be deduced from this chart, thirty years after the highlighted decline, more than 90 percent of the ngrams have gone down further.

212 **"The physicist compares a series of similar facts."** See Franz Boas, "The Study of Geography," *Science* 210S (1887): 137–41.

INDEX

Page numbers in italics indicate charts.